Math to Learn

A Mathematics Handbook

Mary C. Cavanagh

GREAT SOURCE
EDUCATION GROUP
A Houghton Mifflin Company

About the author Mary C. Cavanagh is currently the project coordinator for Math, Science, and Beyond at the Solana Beach School District in Solana Beach, CA.

Acknowledgements We gratefully acknowledge the following teachers and supervisors who helped make *Math to Learn* a reality.

Senior Consultant:

Dr. Marsha W. Lilly
Secondary Math Coordinator
Alief ISD
Houston. Texas

Gale Levow
Mathematics Coordinator
District 8
Bronx, New York

Cynthia M. Petillo
Teacher
Del Prado Elementary (Palm Beach County)
Boca Raton, Florida

Sharon Fields Simpson
Elementary Mathematics Resource Teacher
Cherry Creek Schools
Englewood, Colorado

Kimberly M. Tolbert
Primary Specialist
Alief ISD
Houston, Texas

Marilyn LeRud
Retired Elementary Teacher K-8
Tucson Unified School District
Tucson, Arizona

Louise Muscarella-Daxon
District Mathematics Supervisor
Miami-Dade County Public Schools
Miami, Florida

Sandra Silverman
Coordinator, Elementary Mathematics
San Diego County Office of Education
San Diego, California

Dr. Lon M. Stettler
Instructional Programs Coordinator,
Mathematics, Science, Gifted Education
Cincinnati, Ohio

Darrell Williamson
K-12 Mathematics Consutant
Fulton County
Atlanta, Georgia

Writing: Edward Manfre
Editorial: Carol DeBold; Justine Dunn; Kane Publishing Services, Inc.; Susan Rogalski
Design Management: Richard Spencer
Production Management: Sandra Easton
Design and Production: Bill SMITH STUDIO
Marketing: Lisa Bingen
Illustration credits: see end of index

International Standard Book Number: 0-669-53599-0 (hardcover)
1 2 3 4 5 6 7 8 9 0 QWT 10 09 08 07 06

International Standard Book Number: 0-669-53598-2 (softcover)
1 2 3 4 5 6 7 8 9 0 QWT 10 09 08 07 06

Table of Contents

Multiplication and Division Concepts

Addition and Subtraction with Greater Numbers

Money and Time

Geometry   192

Measurement   206

Graphing, Statistics, and Probability   232

Algebraic Thinking   252

How to Use this Book

This is a math handbook. Use it the same way you use a dictionary or an encyclopedia. You don't need to read it from beginning to end.

Keep this book handy and use it
- when you don't know the meaning of a math word
- if you want to find out how to do a certain kind of problem
- if you just want to know more about a math topic.

What if you need to solve a problem about perimeter? This book can help you even if you don't know what "perimeter" is.

To find out how to use this book to learn about "perimeter," turn the page.

- You can look for "perimeter" in the Glossary of Math Words. The Glossary of Math Words is part of the Yellow Pages which begin on page 313.

Look for the pages that are all yellow! They are near the end of the book.

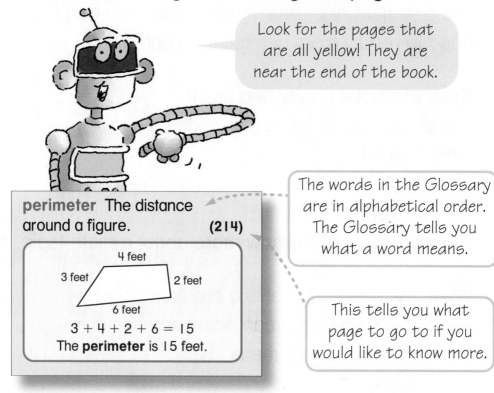

perimeter The distance around a figure.   (214)

4 feet
3 feet
2 feet
6 feet

$3 + 4 + 2 + 6 = 15$

The **perimeter** is 15 feet.

The words in the Glossary are in alphabetical order. The Glossary tells you what a word means.

This tells you what page to go to if you would like to know more.

- You can look for "perimeter" in the Index. The Index is at the very end of the book.

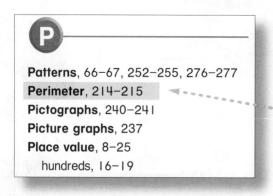

Words in the Index are listed in alphabetical order. Next to each word are page numbers. Perimeter is found on pages 214 and 215.

- You can look for "perimeter" in the Table of Contents. The Table of Contents is at the beginning of the book.

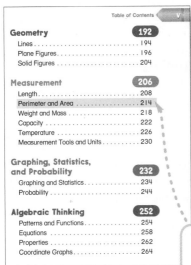

The topics are listed in the Table of Contents in the same order that they are in the book. If you look through, you'll see all the topics that are in this book.

The section on Perimeter and Area starts on page 214.

No matter which way you use to find perimeter, it looks like pages 214 and 215 will help you find out all about it.

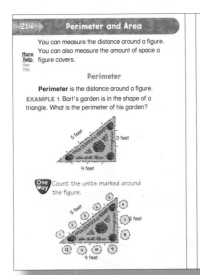

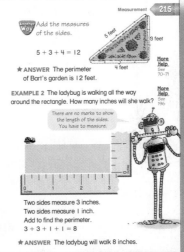

How to Use the Pages

- You can often find more than one way to solve the same problem.

Here's one way.

Here's another way.

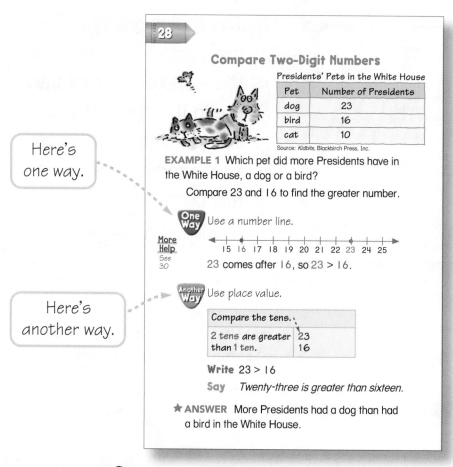

28

Compare Two-Digit Numbers

Presidents' Pets in the White House

Pet	Number of Presidents
dog	23
bird	16
cat	10

Source: *Kidbits*, Blackbirch Press, Inc.

EXAMPLE 1 Which pet did more Presidents have in the White House, a dog or a bird?

Compare 23 and 16 to find the greater number.

One Way Use a number line.

More Help See 30

15 16 17 18 19 20 21 22 23 24 25

23 comes after 16, so 23 > 16.

Another Way Use place value.

Compare the tens.	
2 tens are greater than 1 ten.	23
	16

Write 23 > 16

Say *Twenty-three is greater than sixteen.*

★ **ANSWER** More Presidents had a dog than had a bird in the White House.

Whether you use a number line or place value, you will find out that 23 is greater than 16. When you solve your own problems, choose the way that works best for you.

- **Math Alert** These help keep you from making mistakes.
  This Math Alert shows you how to line up the digits in addition.

MATH ALERT Line Up the Ones to Add

$$\begin{array}{r} 3\ 4 \\ +\ \ \ 5 \\ \hline 3\ 9 \end{array}$$

If you write 5 in the tens column, you will be adding 50, not 5!

- **More Help** These give you page numbers where you can go to find more help.

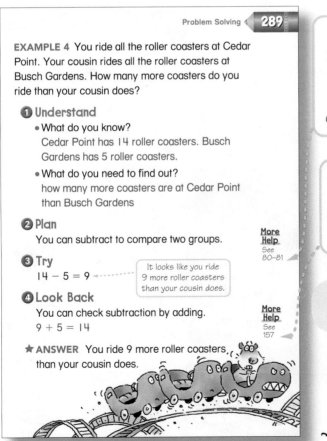

Problem Solving **289**

EXAMPLE 4 You ride all the roller coasters at Cedar Point. Your cousin rides all the roller coasters at Busch Gardens. How many more coasters do you ride than your cousin does?

❶ Understand
- What do you know?
  Cedar Point has 14 roller coasters. Busch Gardens has 5 roller coasters.
- What do you need to find out?
  how many more coasters are at Cedar Point than Busch Gardens

❷ Plan
You can subtract to compare two groups.

More Help See 80–81

❸ Try
$14 - 5 = 9$ — It looks like you ride 9 more roller coasters than your cousin does.

❹ Look Back
You can check subtraction by adding.
$9 + 5 = 14$

More Help See 157

★ **ANSWER** You ride 9 more roller coasters than your cousin does.

On pages 80 and 81, you can find out more about how to use subtraction to compare two numbers.

Page 157 shows you more about using addition to check subtraction.

Have fun using this math handbook!

Numeration

I asked for a cone with some scoops.

Why do we need numbers? Think about a world with no numbers. That will help you see why.

Counting is easier when you see the number patterns.

0　1　2　3　4　5　6　7　8　9
10　11　12　13　14　15　16　17　18　19

Count Each Item Only Once

Move each object as you count it.

1, 2, 3, . . .

Numbers 0 to 9

See	Write	Say
	0	zero
	1	one
	2	two
	3	three
	4	four
	5	five
	6	six
	7	seven
	8	eight
	9	nine

Ten Frames—Numbers 0 to 10

You can use a frame with ten boxes to show numbers from 0 to 10.

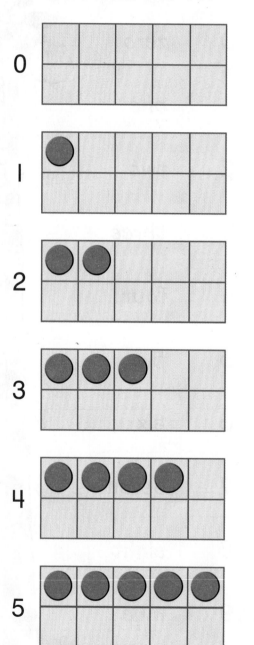

0

1

2

3

4

5

The frame can also help you to see how many more you need to make 5.

If you have 3, you need 2 more to make 5.

Here you can see 5 and some more.

6

7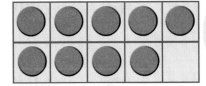

5 and 2 more
is 7.

8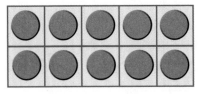

9

5 and 4 more
is 9.

10

I have
many rocks.

So do I.

Did You Know?

A long time ago, some
people used only three
numbers: *one, two,* and *many.*
Source: *The How and Why Wonder Book of
Mathematics,* Wonder Books.

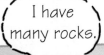

Numbers 10 to 19

Numbers 11 to 19 all have a complete ten frame and then some more.

See	Write	Say
	10	ten
	11	eleven
	12	twelve
	13	thirteen
	14	fourteen
	15	fifteen
	16	sixteen

See	Write	Say
 two ten-frames	17	seventeen
two ten-frames	18	eighteen
two ten-frames	19	nineteen

Numbers 11 to 19 Can Be Tricky

The words for numbers 11 to 19 don't follow a pattern.

Ten, eleven, twelve,... What comes after twelve?

Thirteen. Don't worry, once you get to twenty, counting is easy.

You can write any number using these ten **digits.**

| 0 | 1 | 2 | 3 | 4 | 5 | 6 | 7 | 8 | 9 |

The place that a digit is in tells you how much the digit stands for. This is called its **place value.**

Numbers to 99

You use tens and ones for numbers to 99.
Look for patterns on the chart.

0	1	2	3	4	5	6	7	8	9
10	11	12	13	14	15	16	17	18	19
20	21	22	23	24	25	26	27	28	29
30	31	32	33	34	35	36	37	38	39
40	41	42	43	44	45	46	47	48	49
50	51	52	53	54	55	56	57	58	59
60	61	62	63	64	65	66	67	68	69
70	71	72	73	74	75	76	77	78	79
80	81	82	83	84	85	86	87	88	89
90	91	92	93	94	95	96	97	98	99

Each number in this row has **3 tens**.

More Help
See 10–13

Each number in this column has **6 ones.**

Models for Numbers

Models can help you see place value.
Here are some ways to show fourteen.

 Use ten frames.

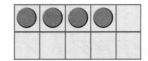

14 has 1 ten and 4 ones.

 Use beansticks and beans.

 Use place-value blocks.

 Use linking cubes.

 Use bundles of sticks.

TEN

Tens

You can make groups of ten when you have many things to count.

10 ones make 1 ten.

3 tens make thirty.

See	Write	Say
	10	ten
	20	twenty
	30	thirty
	40	forty
	50	fifty

See	Write	Say
	60	sixty
	70	seventy
	80	eighty
	90	ninety

Ten more than 90 would be 100. That's one hundred.

More Help
See 16–17

MATH ALERT

Tens May Look Different

I ten

I ten

On its side or on its end, it is still a ten!

Tens and Ones

Sometimes you have extras when you make groups of ten. Then you have tens and ones.

Tens	Ones

Tens	Ones
3	5

More Help
See 14

Think 3 tens 5 ones

Write 35

Say *thirty-five*

Here are more tens and ones.

See	Think	Write
	4 tens 2 ones	42
	2 tens 7 ones	27
	1 ten 6 ones	16
	5 tens 3 ones	53
	3 tens 9 ones	39

Number Words

EXAMPLE 1 There are 52 weeks in a year.
Write 52 in words.

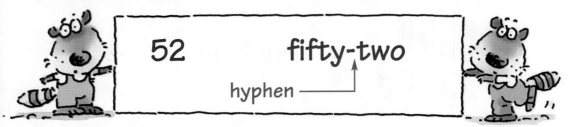

52   fifty-two

hyphen ⟶

⭐ **ANSWER** fifty-two

EXAMPLE 2 Juan's mother is 29 years old.
Write 29 in words.

More Help
See
4–7,
10–11

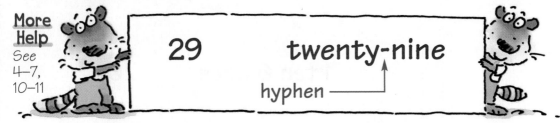

29   twenty-nine

hyphen ⟶

⭐ **ANSWER** twenty-nine

EXAMPLE 3 Sarah has 40 storybooks.
Write 40 in words.

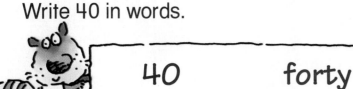

40   forty

You don't need a hyphen for numbers like 20, 30, 40, and so on.

⭐ **ANSWER** forty

Don't Mix Up Tens and Ones

The 6 in 46 means 6 ones.

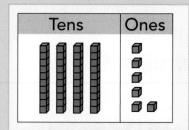

Think 4 tens 6 ones   **Write** 46

                          Say   *forty-six*

The 6 in 64 means 6 tens.

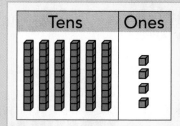

Think 6 tens 4 ones   **Write** 64

                          Say   *sixty-four*

How are you doing?   OK.

But somehow I feel more important over here.

Numbers 100 to 999

You use hundreds, tens, and ones for numbers from 100 to 999.

One Hundred

10 ones make 1 ten.

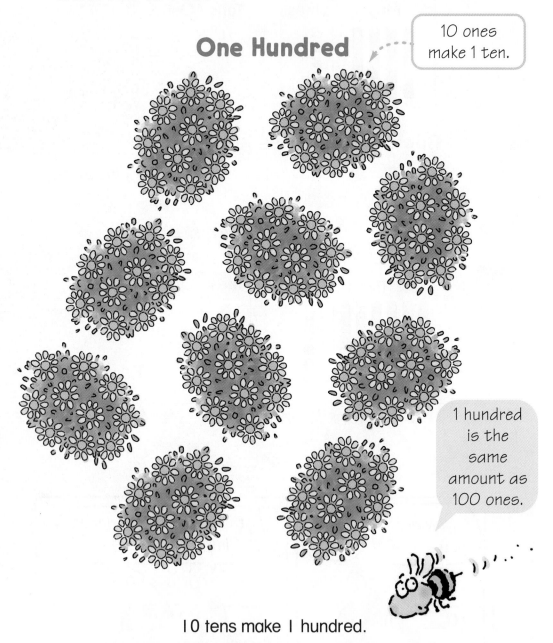

1 hundred is the same amount as 100 ones.

10 tens make 1 hundred.

Here are some ways to show 1 hundred.

 Use beansticks.

These may be easier
to use than
10 bunches of flowers,
but they sure don't
smell as nice!

Another Way Use place-value blocks.

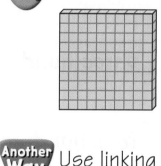

Another Way Use linking cubes.

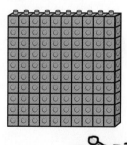

Hundreds

You can count by hundreds.
Just follow the patterns.

See	Write	Say
	100	one hundred
	200	two hundred
	300	three hundred
	400	four hundred
	500	five hundred

See	Write	Say
	600	six hundred
	700	seven hundred
	800	eight hundred
	900	nine hundred

One hundred more than 900 would be 1000. That's one thousand.

More Help
See 24–25

Hundreds, Tens, and Ones

You can show any number from 100 to 999 using hundreds, tens, and ones.

One Way You can use models and place-value charts.

Hundreds	Tens	Ones

Hundreds	Tens	Ones
2	4	5

Think 2 hundreds 4 tens 5 ones

Write 245

Say *two hundred forty-five*

Hundreds	Tens	Ones

Hundreds	Tens	Ones
3	2	7

Think 3 hundreds 2 tens 7 ones

Write 327

Say *three hundred twenty-seven*

You can show the values as addition.
This is called **expanded form**.

Hundreds	Tens	Ones
7	9	2

Think 7 hundreds + 9 tens + 2 ones

Write 700 + 90 + 2

You can add the values of the digits to get the number.

Think 700 + 90 + 2 = 792

Write 792

Say *seven hundred ninety-two*

More
Help
See
14

To say a number with three digits, say the first digit. Then say hundred. Last, say the number made by the last two digits.

792

seven hundred .... ninety-two

MATH ALERT Zero is a Hero

I'm Zero. I keep other digits in the correct places.

40 and 4 are not the same.
They show different amounts.

Tens	Ones
‖‖‖‖	

Tens	Ones
4	0

Think   4 tens 0 ones = 40

Tens	Ones
	▫▫ ▫▫

Tens	Ones
0	4

Think   0 tens 4 ones = 4

406 and 46 are not the same.
They show different amounts.

Hundreds	Tens	Ones

Hundreds	Tens	Ones
4	0	6

Think 4 hundreds + 0 tens + 6 ones

Write 406

Say *four hundred six*

More Help
See
21

Hundreds	Tens	Ones

Hundreds	Tens	Ones
0	4	6

406 is a lot more than 46. What a difference a zero makes!

Think 0 hundreds + 4 tens + 6 ones

Write 46

Say *forty-six*

One Thousand

- When you count to 9, the next number is 10.

- When you count to 99, the next number is 100.

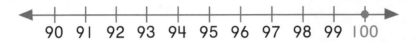

- When you count to 999, the next number is 1000.

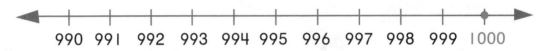

Here are 10 groups with 100 ants in each group.

100 ones make 1 hundred.

1 thousand is the same amount as 1000 ones.

10 hundreds equal 1 thousand.

Here are some ways to show 1000.

 Use place-value blocks.

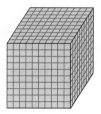

Think 1 thousand

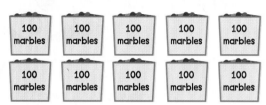 Use 10 boxes with 100 marbles in each.

| 100 marbles | 100 marbles | 100 marbles | 100 marbles | 100 marbles |
| 100 marbles | 100 marbles | 100 marbles | 100 marbles | 100 marbles |

Think 10 hundreds

More Help
See
18–19

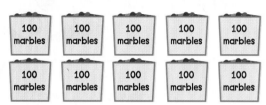 Use 100 boxes with 10 marbles in each.

10	10	10	10	10	10	10	10	10	10
10	10	10	10	10	10	10	10	10	10
10	10	10	10	10	10	10	10	10	10
10	10	10	10	10	10	10	10	10	10
10	10	10	10	10	10	10	10	10	10
10	10	10	10	10	10	10	10	10	10
10	10	10	10	10	10	10	10	10	10
10	10	10	10	10	10	10	10	10	10
10	10	10	10	10	10	10	10	10	10
10	10	10	10	10	10	10	10	10	10

Think 100 tens

Write 1000 or 1,000

Say *one thousand*

You can compare or order numbers by using pictures or models. You can also look at the values of the digits.

Comparing Numbers

You can compare numbers to see if they are the same or to see which is greater or less.

Compare One-Digit Numbers

EXAMPLE 1 Are there more dogs or beds?

> The matching lines show that there is one bed for each dog. The beds are matched one-to-one.

Think   3 dogs, 3 beds

3 and 3 are the same amounts.

Write   3 = 3

Say   *Three equals three.*

★ **ANSWER** The number of dogs equals the number of beds.

EXAMPLE 2 Are there more kittens or puppies?

One Way You can draw a picture. Then draw lines to compare.

5

7

Another Way You can use a number line.

There are 5 kittens and 7 puppies.

> The numbers on a number line are in counting order.

0 1 2 3 4 5 6 7 8 9 10

Think When I count, 7 comes after 5.

7 is **greater than** 5.

When I count, 5 comes before 7.

5 is **less than** 7.

★ **ANSWER** There are more puppies than kittens.

Compare Two-Digit Numbers

Presidents' Pets in the White House

Pet	Number of Presidents
dog	23
bird	16
cat	10

Source: *Kidbits,* Blackbirch Press, Inc.

EXAMPLE 1 Which pet did more Presidents have in the White House, a dog or a bird?

Compare 23 and 16 to find the greater number.

 Use a number line.

More Help See 30

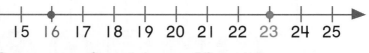

15 16 17 18 19 20 21 22 23 24 25

23 comes after 16, so 23 > 16.

 Use place value.

Compare the tens.	
2 tens are greater than 1 ten.	23 16

Write 23 > 16

Say   *Twenty-three is greater than sixteen.*

★ **ANSWER** More Presidents had a dog than had a bird in the White House.

EXAMPLE 2 Which pet did fewer Presidents have in the White House, a bird or a cat?

Compare 16 and 10 to find which number is less.

More Help
See 30

 One Way Use a number line.

10 comes before 16, so 10 < 16.

 Another Way Use place value.

1 Compare the tens.	
The digits are the same. 1 ten equals 1 ten.	16 10
2 Compare the ones.	
0 is less than 6.	16 10

Write 10 < 16

Say *Ten is less than sixteen.*

★ **ANSWER** Fewer Presidents had a cat than had a bird in the White House.

MATH ALERT

Don't Mix Up > and <

Here's a way to help you write the correct symbol.

1. Place : next to the larger number.
2. Place • next to the smaller number.
3. Connect the dots.

15 ◁ 27

15 is less than 27.

42 ▷ 39

42 is greater than 39.

Compare Three-Digit Numbers

EXAMPLE 1 Compare. 485 ◯ 723

When you see a circle like this, you write the symbol inside of it.

Start at the left.	
Compare the hundreds.	485
400 is less than 700.	723

Think 485 is less than 723.

★ **ANSWER** 485 ⟨<⟩ 723

EXAMPLE 2 In which month did Grade 2 students read more books?

Carmel Creek School
Books Read by Grade 2 Students

Month	Number of Books
January	360
February	342

1 Compare the hundreds.

| The digits are the same. | 3**60** | January |
| 300 equals 300. | 342 | February |

2 Compare the tens.

| | 3**6**0 | January |
| 60 is greater than 40. | 3**4**2 | February |

Write 360 > 342

Say *360 is greater than 342.*

★ **ANSWER** Grade 2 students read more books in January than in February.

Ordering Numbers

It is easier to work with a group of numbers if you put them in order.

Between

EXAMPLE Write the counting number that goes **between** 3 and 5.

3 ___ 5

One Way You can use objects to find the number.

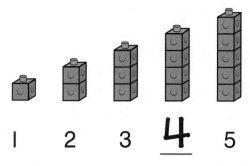

1   2   3   **4**   5

Another Way You can count on a number line or in your head to find the number.

three, four, five

0   1   2   3   4   5   6   7   8   9   10

★ **ANSWER** 4 goes between 3 and 5.

3 **4** 5

Just Before and Just After

EXAMPLE 1 Write the counting number that comes just **after** 26.

26 ___

Count on to find the number just after.

You can use a number chart.

You can also use a number line.

10	11	12	13	14	15	16	17	18	19
20	21	22	23	24	25	26	27	28	29

★ **ANSWER** 27 comes just after 26.

26 **27**

EXAMPLE 2 Write the counting number that comes just **before** 20.

___ 20

Count back to find the number just before.

You can use a number line.

10 11 12 13 14 15 16 17 18 19 20

You can also use a number chart.

★ **ANSWER** 19 comes just before 20.

19 20

Order Three or More Numbers

Sometimes you have to compare more than two numbers. It helps to put the numbers in order.

EXAMPLE 1 Who has the greatest number of cards? Who has the least number of cards?

Put the three numbers in order.

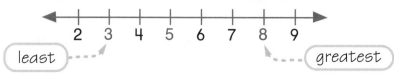

★ **ANSWER** Ida has the greatest number of cards, 8. Luis has the least number of cards, 3.

EXAMPLE 2 Which of these numbers is greatest? Which of these numbers is least?

| 7 | 25 | 4 | 23 | 18 |

The numbers in order are: 4, 7, 18, 23, 25.

★ **ANSWER** 25 is greatest. 4 is least.

Ordinal Numbers

You can use numbers to describe the position or the order of things.

See				
Write	Ⅰst	2nd	3rd	4th
Say	*first*	*second*	*third*	*fourth*

Ⅰ0th	tenth
9th	ninth
8th	eighth
7th	seventh
6th	sixth
5th	fifth
4th	fourth
3rd	third
2nd	second
Ⅰst	first

Odd and Even Numbers

Even numbers
make pairs.

2

> 2 shoes make 1 pair.

Odd numbers
have one left over.

3

> 3 shoes make 1 pair
> with 1 shoe left over.

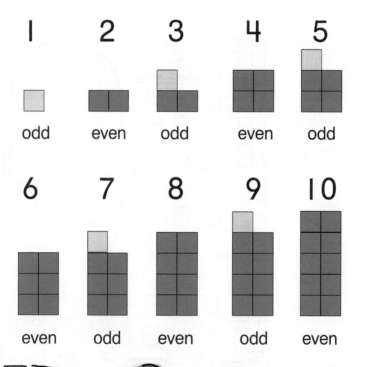

1	2	3	4	5
odd	even	odd	even	odd

6	7	8	9	10
even	odd	even	odd	even

> I have 3 socks.
> That's odd.

Look at the ones place to tell if a number is odd or even.

Even numbers have an even number or zero in the ones place.

| 2 | 4 | 6 | 8 | 10 | 12 | 14 | 16 | 18 | 20 | 22 |

Odd numbers have an odd number in the ones place.

| 1 | 3 | 5 | 7 | 9 | 11 | 13 | 15 | 17 | 19 | 21 |

This works for any size number. The number 972 is even, because the 2 in the ones place is even.

EXAMPLE Find the even and odd numbers on the chart.

Hundred Chart

1	2	3	4	5	6	7	8	9	10
11	12	13	14	15	16	17	18	19	20
21	22	23	24	25	26	27	28	29	30
31	32	33	34	35	36	37	38	39	40
41	42	43	44	45	46	47	48	49	50

★ **ANSWER** Even numbers are in the green boxes. Odd numbers are in the yellow boxes.

Many times you do not need to know exactly how many. You may only need to know *about* how many. That is when you **estimate**.

There are about 50 children ahead of us.

Yes, but there are about 300 children behind us!

Estimates are not exact. So, you can have more than one good estimate. Some estimates are better than others.

EXAMPLE 1 Are there about 10 tickets or about 30 tickets? Estimate to decide.

Think I know what 10 looks like. This looks like a little more than 10.

10 is a better estimate than 30. Sometimes 30 might be OK. It would be a better estimate than 100 or 1000!

★ **ANSWER** There are about 10 tickets.

Sometimes you can use one amount to estimate another amount.

EXAMPLE 2 Estimate how many marbles are in the large jar.

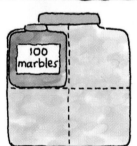

Think The large jar is as wide as two small jars. It is as tall as two small jars.

Think Jars are not flat.

> The large jar is as deep as two small jars. There are many marbles I can't see.

It would take about 8 small jars to fill 1 large jar.

> 8 hundreds = 800
> 800 is probably a good estimate.

★ **ANSWER** There are about 800 marbles in the large jar.

Rounding

You can **round a number** to make it easier to work with. When you **round up,** you get a higher number. When you **round down,** you get a lower number.

Round to the Nearest Ten

More
Help
See
10–11

A number line can help you round to the nearest ten.

EXAMPLE 1 Round 18 to the nearest ten.

Think • What is the nearest ten before 18?
• What is the nearest ten after 18?
• Which ten is closer to 18?

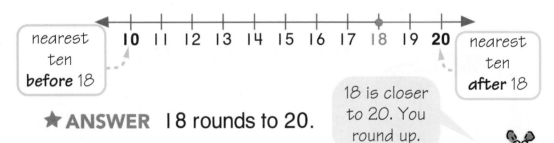

nearest ten **before** 18

10 11 12 13 14 15 16 17 18 19 **20**

nearest ten **after** 18

★ **ANSWER** 18 rounds to 20.

18 is closer to 20. You round up.

EXAMPLE 2 Round 43 to the nearest ten.

40 41 42 43 44 45 46 47 48 49 **50**

★ **ANSWER** 43 rounds to 40.

43 is closer to 40. You round down.

EXAMPLE 3 Round 35 to the nearest ten.

Here is a rule.

When a number has 5 ones, round up to the next ten.

30 31 32 33 34 35 36 37 38 39 **40**

★ **ANSWER** 35 rounds to 40.

Round to the Nearest Hundred

More Help
See 18–19

EXAMPLE Round 182 to the nearest hundred.

Think Is 182 closer to 100 or to 200?

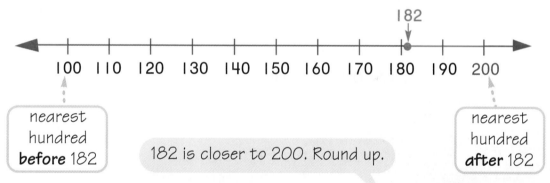

182

100 110 120 130 140 150 160 170 180 190 200

nearest hundred **before** 182

182 is closer to 200. Round up.

nearest hundred **after** 182

★ **ANSWER** 182 rounds to 200.

A **fraction** is a number that stands for part of something.

You can use a fraction to tell how much of the flag is green.

The flag has 3 equal parts.

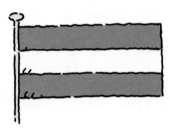

2 of the parts are green.

The **numerator** tells the number of green parts. → $\dfrac{2}{3}$ ← The **denominator** tells the total number of equal parts in the whole.

So, $\dfrac{2}{3}$ of the flag is green.

MATH ALERT

Equal Parts Make Fractions

There must be **equal parts** of the whole to make a fraction.

That's not fair. You did not make equal parts.

| 2 equal parts | 3 equal parts | 4 equal parts |

Fraction of a Whole

A **fraction** can name part of one thing.

Halves

2 equal parts are **halves**.

Each part is **one half**.

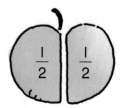

Halves do not all look the same.

One half of each shape is red.

  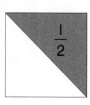 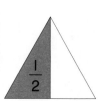

Think   1 out of 2 equal parts

Write $\dfrac{1}{2}$

Say   *one half*

Thirds

3 equal parts
are **thirds**.

Each part is
one third.

Thirds do not all look the same.

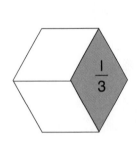

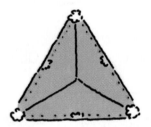

One third of each shape is orange.

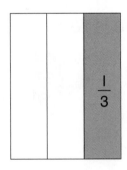

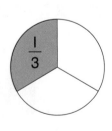

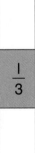

Think   1 out of 3 equal parts

Write $\frac{1}{3}$

Say   *one third*

Here are some other ways to use thirds.

EXAMPLE 1 What fraction of the spinner is yellow?

Think There are 3 equal parts.
2 out of 3 equal parts
are yellow.

Write $\dfrac{2}{3}$

Say *two thirds*

★ **ANSWER** $\dfrac{2}{3}$ of the spinner is yellow.

$\dfrac{1}{3}$ *of the spinner is not yellow.*

EXAMPLE 2 Ed walks to school. What fraction of the way has he walked?

Ed's house   Ed   School

Think There are 3 equal parts.
Ed has walked 1 part.

Write $\dfrac{1}{3}$

Say *one third*

★ **ANSWER** Ed has walked $\dfrac{1}{3}$ of the way.

Fourths

4 equal parts are **fourths**.

Each part is **one fourth**.

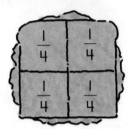

Fourths do not all look the same.

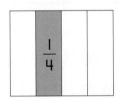

One fourth of each shape is green.

More Help
See 168

Think I out of 4 equal parts

Write $\frac{1}{4}$

Say one fourth

You may also say "one quarter" for one fourth. Remember that a quarter is $\frac{1}{4}$ of a dollar.

Here are some other ways to use fourths.

EXAMPLE 1 What fraction of the pizza is plain?

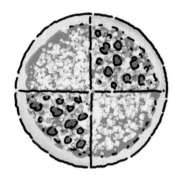

Think There are 4 equal parts. 2 of 4 equal parts are plain.

Write $\frac{2}{4}$

Say *two fourths*

★ **ANSWER** $\frac{2}{4}$ of the pizza is plain.

You could also say that $\frac{1}{2}$ of the pizza is plain.

More Help See 51

EXAMPLE 2 What fraction of the sandwich is left?

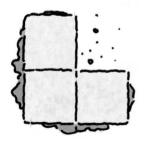

Think There were 4 equal parts. 3 parts are left.

Write $\frac{3}{4}$

Say *three fourths*

★ **ANSWER** $\frac{3}{4}$ of the sandwich is left.

Fraction of a Group

You can use a fraction to name part of a group.

EXAMPLE 1 Two of the three balloons are yellow. What fraction of the balloons are yellow?

Think 2 out of 3 are yellow.

Write $\frac{2}{3}$

Say *two thirds*

★ **ANSWER** $\frac{2}{3}$ of the balloons are yellow.

$\frac{1}{3}$ of the balloons are purple.

EXAMPLE 2 What fraction of the hats are red?

Think 1 out of 4 is red.

Write $\frac{1}{4}$

Say *one fourth*

★ **ANSWER** $\frac{1}{4}$ of the hats are red.

$\frac{3}{4}$ of the hats are not red.

EXAMPLE 3 What fraction of the balls are yellow?

Think 2 out of 5 are yellow.

Write $\dfrac{2}{5}$

$\dfrac{3}{5}$ of the balls are not yellow.

Say *two fifths*

★ **ANSWER** $\dfrac{2}{5}$ of the balls are yellow.

EXAMPLE 4 What fraction of the kites are red?

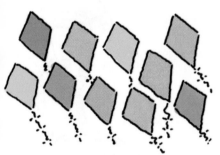

Think I out of 10 is red.

Write $\dfrac{1}{10}$

$\dfrac{9}{10}$ of the kites are not red.

Say *one tenth*

★ **ANSWER** $\dfrac{1}{10}$ of the kites are red.

Here are other fractions you might see.

$\dfrac{1}{6}$

one sixth

$\dfrac{3}{5}$

three fifths

$\dfrac{5}{8}$

five eighths

$\dfrac{2}{10}$

two tenths

Comparing Fractions

You can compare fractions to see if they are the same or to see which is greater or which is less.

Unequal Fractions

Fractions can show different amounts.

EXAMPLE 1 Compare. $\frac{1}{2} \bigcirc \frac{1}{3}$

Use models.

| $\frac{1}{2}$ | |
| $\frac{1}{3}$ | | |

One half is greater than one third.

★ **ANSWER** $\frac{1}{2} \gtrdot \frac{1}{3}$

EXAMPLE 2 Compare. $\frac{1}{4} \bigcirc \frac{1}{3}$

Use models.

| $\frac{1}{4}$ | | | |
| $\frac{1}{3}$ | | |

One fourth is less than one third.

★ **ANSWER** $\frac{1}{4} \lessdot \frac{1}{3}$

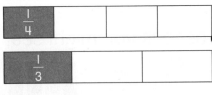

4 is greater than 3, but $\frac{1}{4}$ is less than $\frac{1}{3}$. That makes sense. The more parts there are, the smaller the parts are.

Equal Fractions

Different fractions can show the same amount.

EXAMPLE Compare. $\frac{2}{4} \bigcirc \frac{1}{2}$

Use models.

Two fourths equals one half.

★ **ANSWER** $\frac{2}{4} \bigcirc\!\!= \frac{1}{2}$

Use the Same-Size Whole

When you compare fractions, make sure the wholes or the groups are the same size.

$\frac{1}{4}$ of a large pie does not equal $\frac{1}{4}$ of a small pie..

My fourth is bigger than your fourth.

Addition and Subtraction
Concepts and Facts

Mmmmmm.

You can always figure out addition and subtraction facts. But if you learn these facts by heart, you can save a lot of time.

We need plates for 3 girls and 4 boys. Let's see, 3 plus 4 is …mmmmmm.

Addition

Here are some reasons you might use **addition**.

- One group joins another group. Add to find how many in all.

3 ducks in a pond.

1 more duck comes to the pond.

Add **3** and **1**.
There are **4** ducks in all.

- There are two parts. Add to find the whole amount.

5 gray bunnies.   **3** white bunnies.

Add **5** and **3**.
There are
8 bunnies in all.

● You know how many *more*. Add to find how many.

 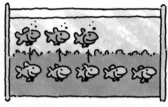

Add **3** and **2**.
There are **5**
yellow fish.

3 blue fish.   **2** more yellow fish
than blue fish.

Addition Words and Symbols

The numbers to **add** are **addends**.

The total number is the **sum**.

The **plus sign** tells you to add.

The **equal sign** means *is equal to*
or *is the same as*.

See

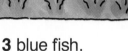

2 red bears.
I green bear.
3 bears in all.

These are
two ways to
show the
same
addition.

Write

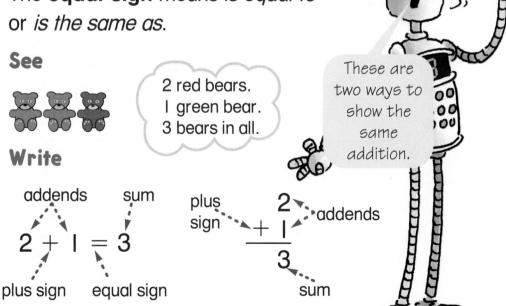

addends   sum

$2 + 1 = 3$

plus sign   equal sign

plus
sign

$$\begin{array}{r} 2 \\ +\ 1 \\ \hline 3 \end{array}$$

addends

sum

Say *Two plus one equals three.*

Learning to Add

There are many ways to find a sum.

Use Models to Add

EXAMPLE 2 + 4 = ■

 Use linking cubes. Find how many in all.

More
Help
See 9

Show 2. Add 4 more.   There are 6 in all.

 Use a part-part-whole mat.
Find the whole amount.

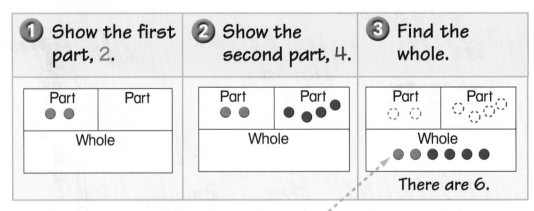

① Show the first part, 2.	② Show the second part, 4.	③ Find the whole.
Part Part	Part Part	Part Part
Whole	Whole	Whole
		There are 6.

★ **ANSWER** 2 + 4 = 6

If you don't have counters like these, use something else—like crayons.

Draw a Picture to Add

Draw to show each addend.

Count how many.

EXAMPLE 1 4 + 3 = ■

⭐ **ANSWER** 4 + 3 = 7

EXAMPLE 2 2 + 5 = ■

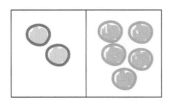

2 + 5 = 7

⭐ **ANSWER** 2 + 5 = 7

You don't need to draw fancy pictures.

Addition Fact Strategies

You often use a clever plan, or **strategy**, when you play a game like checkers. You can learn strategies to help you add.

Turn-Around Facts in Addition

You can add in any order.
You will get the same sum.

More
Help
See
262

If you know	Then you also know
$1+3=4$	$3+1=4$
$7+3=10$	$3+7=10$
$\begin{array}{r} 4 \\ +\ 1 \\ \hline 5 \end{array}$	$\begin{array}{r} 1 \\ +\ 4 \\ \hline 5 \end{array}$

I like to chase my tail...

so I turn around often!

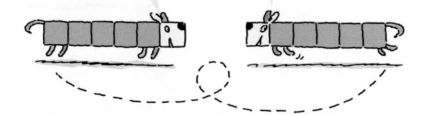

Adding Zero

When you add zero to a number, the sum is that same number.

EXAMPLE $3 + 0 = $ ■

3 hens 　　　　　　　　　　0 hens

Think 3 hens and no hens is still 3 hens.

★ **ANSWER** $3 + 0 = 3$

More
Help
See
262

Now you know how to add zero. So you know all these addition facts.

More
Help
See
58,
262

$0 + 0 = 0$	$0 + 0 = 0$
$1 + 0 = 1$	$0 + 1 = 1$
$2 + 0 = 2$	$0 + 2 = 2$
$3 + 0 = 3$	$0 + 3 = 3$
$4 + 0 = 4$	$0 + 4 = 4$
$5 + 0 = 5$	$0 + 5 = 5$
$6 + 0 = 6$	$0 + 6 = 6$
$7 + 0 = 7$	$0 + 7 = 7$
$8 + 0 = 8$	$0 + 8 = 8$
$9 + 0 = 9$	$0 + 9 = 9$

Count On to Add

More Help
See 26–27, 58

You can use your counting skills to find a sum.

One Way Put the greater number in your head. Count on.

See	Think	Write
6 + 2 = ■		6 + 2 = 8
9 + 1 = ■		9 + 1 = 10
3 + 8 = ■		3 + 8 = 11

Remember, the sum of 8 and 3 is the same as the sum of 3 and 8.

Use a number line. Count on from the greater number.

More Help
See 26–27

EXAMPLE $5 + 3 = $ ■

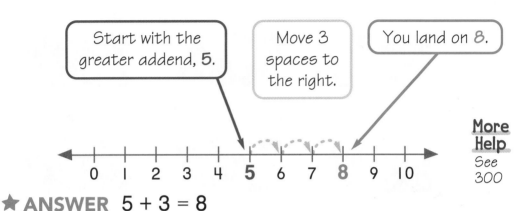

Start with the greater addend, **5**.

Move 3 spaces to the right.

You land on **8**.

0 1 2 3 4 **5** 6 7 **8** 9 10

More Help
See 300

⭐ **ANSWER** $5 + 3 = 8$

MATH ALERT **When You Use a Number Line to Count On, Count Each "Jump"**

Think about moving your marker on a game board.

Adding Doubles

A **doubles** fact has two addends that are the same.

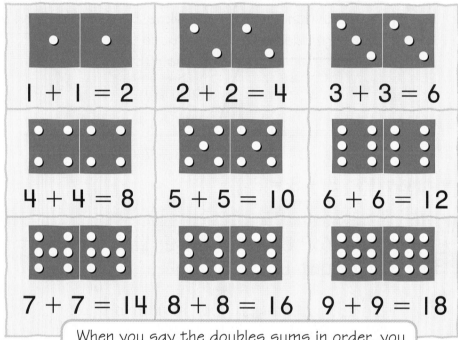

1 + 1 = 2	2 + 2 = 4	3 + 3 = 6
4 + 4 = 8	5 + 5 = 10	6 + 6 = 12
7 + 7 = 14	8 + 8 = 16	9 + 9 = 18

More Help
See 97

When you say the doubles sums in order, you count by twos. 2, 4, 6, 8, and so on.

Doubles are everywhere.

6 + 6	7 + 7	5 + 5
12 cans	14 days	10 fingers

Doubles Plus 1

A **double plus 1** has an addend that is 1 more than the other addend.

Double	Double Plus 1
$2 + 2 = 4$	$2 + 3 = 5$
	2 + 2 and 1 more
$5 + 5 = 10$	$5 + 6 = 11$
	5 + 5 and 1 more
$8 + 8 = 16$	$8 + 9 = 17$
	8 + 8 and 1 more

More Help
See 58, 60

Since I can add in any order, I also know that $9 + 8 = 17$.

Make a Ten to Add

To find a sum greater than 10, think about filling a ten frame.

EXAMPLE 1 9 + 2 = ■

More Help
See 4–7, 67

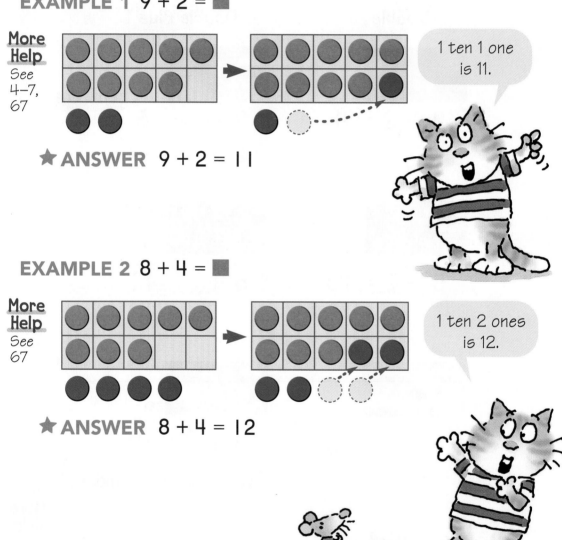

1 ten 1 one is 11.

★ **ANSWER** 9 + 2 = 11

EXAMPLE 2 8 + 4 = ■

More Help
See 67

1 ten 2 ones is 12.

★ **ANSWER** 8 + 4 = 12

Add to 9

EXAMPLE $9 + 4 = \blacksquare$

Make a ten when one addend is 9.

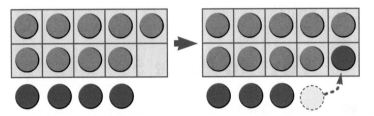

More Help
See 55

1 ten plus 3 ones = 13

★ **ANSWER** $9 + 4 = 13$

Here are some facts with 9 as an addend.

Look at the pattern.

> When one addend is 9, the ones digit in the sum is 1 less than the other addend.

More Help
See 58, 59

$9 + 1 = 10$	$1 + 9 = 10$
$9 + 2 = 11$	$2 + 9 = 11$
$9 + 3 = 12$	$3 + 9 = 12$
$9 + 4 = 13$	$4 + 9 = 13$
$9 + 5 = 14$	$5 + 9 = 14$
$9 + 6 = 15$	$6 + 9 = 15$
$9 + 7 = 16$	$7 + 9 = 16$
$9 + 8 = 17$	$8 + 9 = 17$
$9 + 9 = 18$	$9 + 9 = 18$

2 is 1 less than 3.

This doesn't work when the other addend is zero.

Use Patterns to Make Sums

There can be more than one way to make the same sum. Look for patterns. Start with 0 as one of the addends.

Ways to Make 5

Here is a way to show pairs of numbers with a sum of 5.

The red addends show a 1-more pattern.

The blue addends show a 1-less pattern.

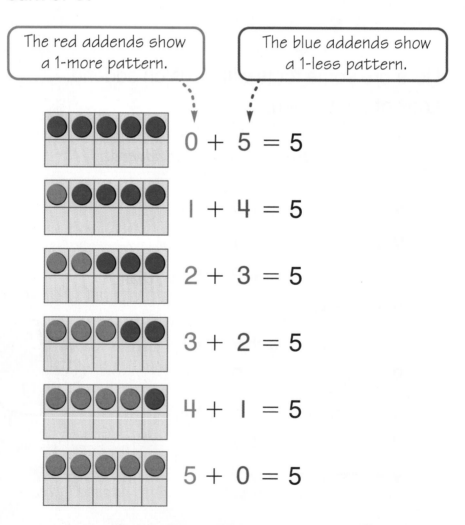

$0 + 5 = 5$

$1 + 4 = 5$

$2 + 3 = 5$

$3 + 2 = 5$

$4 + 1 = 5$

$5 + 0 = 5$

Ways to Make 10

Here is a way to show pairs of numbers with a sum of 10.

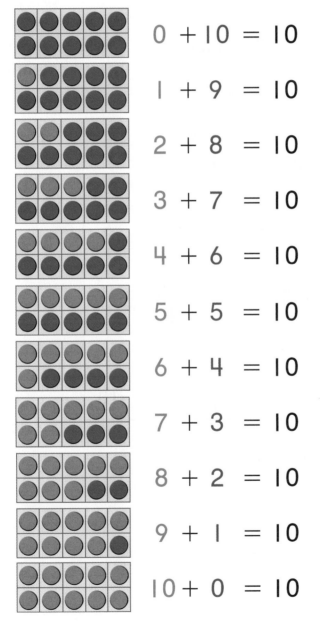

$0 + 10 = 10$

$1 + 9 = 10$

$2 + 8 = 10$

$3 + 7 = 10$

$4 + 6 = 10$

$5 + 5 = 10$

$6 + 4 = 10$

$7 + 3 = 10$

$8 + 2 = 10$

$9 + 1 = 10$

$10 + 0 = 10$

I have $6. I need $4 more and then I'll have $10.

I have $4. I need $6 more before I have $10.

Use an Addition Table

You can find sums on an addition table.

EXAMPLE 1 $3 + 4 = \blacksquare$

1 Find the row for 3.	**2** Find the column for 4.	**3** The sum is where the row and column meet.

+	0	1	2	3	4	5
0	0	1	2	3	4	5
1	1	2	3	4	5	6
2	2	3	4	5	6	7
3	3	4	5	6	⑦	8
4	4	5	6	7	8	9
5	5	6	7	8	9	10

★ **ANSWER** $3 + 4 = 7$

EXAMPLE 2 Jan has 7 blue buttons and 5 yellow buttons. How many buttons does she have?

To solve the problem, add. $7 + 5 = \blacksquare$

Find where the row for 7 and the column for 5 meet on the table.

+	0	1	2	3	4	5	6	7	8	9
0	0	1	2	3	4	5	6	7	8	9
1	1	2	3	4	5	6	7	8	9	10
2	2	3	4	5	6	7	8	9	10	11
3	3	4	5	6	7	8	9	10	11	12
4	4	5	6	7	8	9	10	11	12	13
5	5	6	7	8	9	10	11	12	13	14
6	6	7	8	9	10	11	12	13	14	15
7	7	8	9	10	11	12	13	14	15	16
8	8	9	10	11	12	13	14	15	16	17
9	9	10	11	12	13	14	15	16	17	18

★ **ANSWER** Jan has 12 buttons.

$7 + 5 = 12$

Adding 3 or More Addends

You can group numbers and add in any order.
You will get the same sum.

EXAMPLE 1 How many points?
To solve the problem, add.

$5 + 4 + 1 = \blacksquare$

You can only add two
numbers at a time.

One Way You can add down.

More Help
See 262

Write	**Think**
5	⑤ 9
4	④
+ 1	+ 1
	10

> Group the top two
> addends and add.
> $5 + 4 = 9$

> Then add the third
> addend to find the sum.
> $9 + 1 = 10$

Another Way You can add up.

Write	**Think**
5	5
4	④ 5
+ 1	+ ①
	10

> Start at the bottom of the
> column. Group and then add.
> $1 + 4 = 5$
> $5 + 5 = 10$

★ **ANSWER** 10 points

Look for doubles or other easy facts.

EXAMPLE 2 $3 + 5 + 3 = \blacksquare$

Think

$3 + 5 + 3$

$6 + 5 = 11$

> 3 + 3 is easy because it is a double. Add 5 to the 6 to find the sum.

More Help
See 62

★ **ANSWER** $3 + 5 + 3 = 11$

Look for numbers that make 10.

EXAMPLE 3 $2 + 3 + 7 = \blacksquare$

Think

$2 + 3 + 7$

$10 + 2 = 12$

More Help
See 67

★ **ANSWER** $2 + 3 + 7 = 12$

> You can add more than 3 numbers.
> $3 + 2 + 3 + 8 + 1 = 17$
> First I made a 10 with 2 + 8.
> Then I added the double 3 + 3 = 6.
> 10 + 6 makes 16.
> Adding 1 to 16 was the easiest part.

Here are some reasons you might use
subtraction.

- Part of a group is taken away. Subtract to find
  how many are left.

4 ducks in
a pond.

1 duck leaves
the pond.

Subtract **1** from **4**.
There are **3** ducks
left in the pond.

- You want to compare. Subtract to find how
  many more or fewer.

3 white
bunnies.

5 gray bunnies.

Subtract **3** from **5**.
There are **2** more
gray bunnies than
white bunnies.

- You know the whole and you know one part. Subtract to find the missing part.

More Help See 84– 85

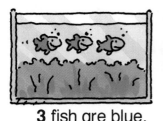

3 fish are blue.　　5 fish in all.

Subtract **3** from **5**.
2 fish are *not* blue.

Subtraction Words and Symbols

You can **subtract** one number from another.

The number you get is the **difference**.

The **minus sign** tells you to subtract.

The **equal sign** means *is equal to* or *is the same as*.

See

3 bears.
2 fall down.
1 left standing.

Write

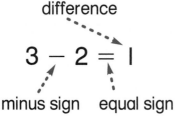

difference

$3 - 2 = 1$

minus sign　equal sign

minus sign

$$\begin{array}{r} 3 \\ -\ 2 \\ \hline 1 \end{array}$$

difference

Say *Three minus two equals one.*

Learning to Subtract

More Help
See 9

There are many ways to find a difference.

Use Models to Subtract

EXAMPLE $5 - 2 = \blacksquare$

One Way Use linking cubes. Find how many are left.

More Help
See 56

Show 5.

Take away 2.
There are 3 left.

Another Way Use a part-part-whole mat.
Find the missing part.

1 Show the whole, 5.	**2** Make the part for 2.	**3** Make the missing part, 3.
Whole ●●●●●	Whole ○○●●●	Whole ○○○○○
Part Part	Part ●● Part	Part ●● Part ●●●

★ **ANSWER** $5 - 2 = 3$

Draw a Picture to Subtract

Draw how many you start with.

Cross out the ones you subtract.

Count how many are left.

EXAMPLE 1 6 − 4 = ■

★ **ANSWER** 6 − 4 = 2

> You can just draw dots or lines. Then cross out to subtract.

EXAMPLE 2 9 − 6 = ■

★ **ANSWER** 9 − 6 = 3

MATH ALERT

Look at Pictures Carefully

Sometimes you have to decide what a picture shows.

This picture shows addition.

5 + 3 = 8

This picture shows subtraction.

8 − 3 = 5

Subtraction Fact Strategies

There are many **strategies** you can use to find a difference.

Count Back to Subtract

One Way Put the start number in your head. Count back.

It helps to know how to count backward from any number.
10, 9, 8, 7, 6, 5, 4, 3, 2, 1

EXAMPLE 1 $7 - 2 = \blacksquare$

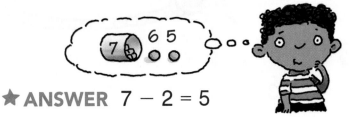

★ **ANSWER** $7 - 2 = 5$

Another Way Find the start number on a number line. Count back.

EXAMPLE 2 $8 - 3 = \blacksquare$

Move 3 spaces back.

More Help
See 60–61

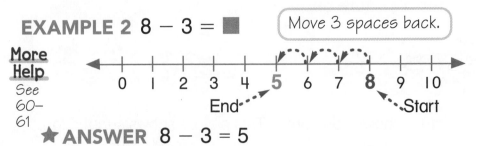

★ **ANSWER** $8 - 3 = 5$

Count Up to Subtract

Sometimes you can count up from the number you subtract.

EXAMPLE 1 8 − 5 = ■

Think 8 is *how many more* than 5?
Start with 5. Count up to 8.
That's 3 jumps.

★ ANSWER 8 − 5 = 3

> Count up to subtract when the two numbers are close to each other.

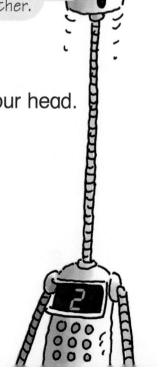

EXAMPLE 2 11 − 9 = ■

Start with 9. Count up to 11 in your head.

9 (10) (11)

That's 2 counts.

★ ANSWER 11 − 9 = 2

Zeros in Subtraction

- When you subtract zero from a number, the difference is that same number.

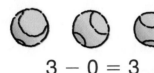

$3 - 0 = 3$

> 3 balls. None are taken away. There are still 3 balls.

- When you subtract a number from itself, the difference is zero.

$3 - 3 = 0$

> 3 balls. 3 are taken away. No balls are left.

Now you know all these facts.

> And don't forget that $0 - 0 = 0$.

$1 - 0 = 1$	$1 - 1 = 0$
$2 - 0 = 2$	$2 - 2 = 0$
$3 - 0 = 3$	$3 - 3 = 0$
$4 - 0 = 4$	$4 - 4 = 0$
$5 - 0 = 5$	$5 - 5 = 0$
$6 - 0 = 6$	$6 - 6 = 0$
$7 - 0 = 7$	$7 - 7 = 0$
$8 - 0 = 8$	$8 - 8 = 0$
$9 - 0 = 9$	$9 - 9 = 0$

> None are taken away.

> All are taken away, so the difference is 0.

Subtracting 1 or 2

- When you subtract 1 from a number, the difference is 1 less than the number.

$$7 - 1 = 6$$

6 is 1 less than 7.

- When you subtract 2 from a number, the difference is 2 less than the number.

$$7 - 2 = 5$$

5 is 2 less than 7.

Now you know these facts.

$2 - 1 = 1$	$2 - 2 = 0$
$3 - 1 = 2$	$3 - 2 = 1$
$4 - 1 = 3$	$4 - 2 = 2$
$5 - 1 = 4$	$5 - 2 = 3$
$6 - 1 = 5$	$6 - 2 = 4$
$7 - 1 = 6$	$7 - 2 = 5$
$8 - 1 = 7$	$8 - 2 = 6$
$9 - 1 = 8$	$9 - 2 = 7$
$10 - 1 = 9$	$10 - 2 = 8$

Using Subtraction to Compare

You can subtract to compare two amounts.

EXAMPLE 1 How many more red counters are there than yellow counters?

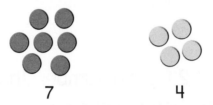

7   4

Line up the groups to compare.

See   **Write** $7 - 4 = 3$

3 more red
than yellow

★ **ANSWER** There are 3 more red counters than yellow counters.

EXAMPLE 2 How many cubes shorter is the green tower?

See   **Write**

$$\begin{array}{r} 5 \\ -\ 3 \\ \hline 2 \end{array}$$

★ **ANSWER** The green tower is 2 cubes shorter than the yellow tower.

> You also know that the yellow tower is 2 cubes taller than the green tower.

EXAMPLE 3 In Karif's school orchestra, 11 students play the clarinet and 6 students play the trumpet. How many more students play the clarinet than play the trumpet?

Subtract to compare 11 to 6.

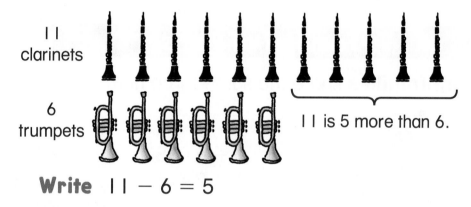

11 clarinets

6 trumpets

11 is 5 more than 6.

Write $11 - 6 = 5$

★ **ANSWER** There are 5 more students who play clarinet than play trumpet.

Did You Know?

Musicians who play similar instruments sit together in an orchestra. This gives the best blending of sounds.

Knowing a related addition fact can help you subtract.

$$4 + 5 = 9 \qquad 9 - 5 = 4$$

Families of Facts

Fact families share the same three numbers.

$$6 + 2 = 8$$
$$2 + 6 = 8$$
$$8 - 2 = 6$$
$$8 - 6 = 2$$

This fact family shares the numbers 2, 6, and 8.

Doubles have only two facts in their family.

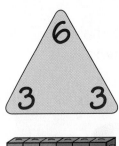

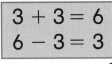

$$3 + 3 = 6$$
$$6 - 3 = 3$$

Sometimes you have to find all the facts in a family.

EXAMPLE 1 Complete the fact family.

$$5 + 1 = \blacksquare \qquad 1 + 5 = \blacksquare$$
$$\blacksquare - 5 = 1 \qquad \blacksquare - 1 = 5$$

Complete the easiest fact first. Use that number to complete the fact family.

Think $5 + 1 = 6$

> 6 also works for the other three facts.

★ **ANSWER**

$$5 + 1 = 6 \qquad 1 + 5 = 6$$
$$6 - 5 = 1 \qquad 6 - 1 = 5$$

EXAMPLE 2 Use these numbers to write a family of facts.

$$9 \qquad 2 \qquad 7$$

Use the greatest number as the sum.

Think $\blacksquare + \blacksquare = 9$

$$2 + 7 = 9$$

Then write the other facts.

★ **ANSWER**

$$2 + 7 = 9 \qquad 7 + 2 = 9$$
$$9 - 7 = 2 \qquad 9 - 2 = 7$$

Missing Addends

EXAMPLE 1 Leroy sold 11 balloons. 8 balloons were orange. The rest were yellow. How many yellow balloons did Leroy sell?

> You need to find this **missing addend**.

More Help
See 73

Think 8 + ■ = 11

orange balloons

yellow balloons

balloons sold

One Way Use models and a part-part-whole mat.

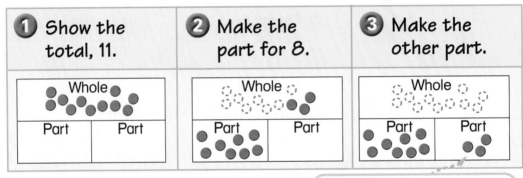

① Show the total, 11.	② Make the part for 8.	③ Make the other part.

> The missing addend is 3.

Another Way Use a fact you know.

Think 8 plus *what* equals 11?

$$8 + ■ = 11$$

More Help
See 77, 82–83

You can use the addition fact $8 + 3 = 11$.

You can use the subtraction fact

$11 - 8 = 3$.

★ **ANSWER** Leroy sold 3 yellow balloons.

EXAMPLE 2 Tara had some balloons. Don gave her 9 more. Now she has 13 balloons. How many balloons did Tara have to start with?

More Help
See 74

Think ■ + 9 = 13

One Way Use models and a part-part-whole mat.

1 Show the total, 13.	**2** Make the part for 9.	**3** Make the other part.
Whole	Whole	Whole
Part Part	Part Part	Part Part

The missing addend is 4.

Another Way Use a fact you know.

Think *What* plus 9 equals 13?

You can use the addition fact 4 + 9 = 13.

You can use the subtraction fact 13 − 9 = 4.

★ **ANSWER** Tara had 4 balloons to start with.

More Help
See 62, 82–83

Using Doubles in Subtraction

Use addition doubles to help you subtract.

$1 + 1 = 2$
$2 - 1 = 1$

$2 + 2 = 4$
$4 - 2 = 2$

$3 + 3 = 6$
$6 - 3 = 3$

$4 + 4 = 8$
$8 - 4 = 4$

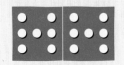

$5 + 5 = 10$
$10 - 5 = 5$

$6 + 6 = 12$
$12 - 6 = 6$

$7 + 7 = 14$
$14 - 7 = 7$

$8 + 8 = 16$
$16 - 8 = 8$

$9 + 9 = 18$
$18 - 9 = 9$

We are the only two facts in our family.

$$\begin{array}{r} 8 \\ + 8 \\ \hline 16 \end{array}$$

$$\begin{array}{r} 16 \\ - 8 \\ \hline 8 \end{array}$$

Subtract from Ten

More Help
See 67, 82–83

Sums for ten can help you subtract from 10.

If you know	Then you also know	
$10 + 0 = 10$	$10 - 0 = 10$	$10 - 10 = 0$
$9 + 1 = 10$	$10 - 1 = 9$	$10 - 9 = 1$
$8 + 2 = 10$	$10 - 2 = 8$	$10 - 8 = 2$
$7 + 3 = 10$	$10 - 3 = 7$	$10 - 7 = 3$
$6 + 4 = 10$	$10 - 4 = 6$	$10 - 6 = 4$
$5 + 5 = 10$	$10 - 5 = 5$	

More Help

See 82–84

Use an Addition Table to Subtract

To subtract, think about a related addition fact.

EXAMPLE 1 $7 - 3 = \blacksquare$

Think $3 + \blacksquare = 7$

① Find the row for 3.	② Find the 7 in that row.	③ The difference is the number at the top of that column.

+	0	1	2	3	④	5
0	0	1	2	3	4	5
1	1	2	3	4	5	6
2	2	3	4	5	6	7
3	3	4	5	6	7	8
4	4	5	6	7	8	9
5	5	6	7	8	9	10

★ **ANSWER** $7 - 3 = 4$

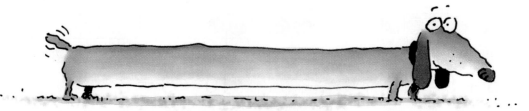

EXAMPLE 2 Luis had 13 cookies. He gave 5 away. How many cookies does he have left?

Write $13 - 5 = \blacksquare$ **Think** $5 + \blacksquare = 13$

① Find the row for 5.	② Find the 13 in that row.	③ The difference is the number at the top of that column.

+	0	1	2	3	4	5	6	7	⑧	9
0	0	1	2	3	4	5	6	7	8	9
1	1	2	3	4	5	6	7	8	9	10
2	2	3	4	5	6	7	8	9	10	11
3	3	4	5	6	7	8	9	10	11	12
4	4	5	6	7	8	9	10	11	12	13
5	5	6	7	8	9	10	11	12	13	14
6	6	7	8	9	10	11	12	13	14	15
7	7	8	9	10	11	12	13	14	15	16
8	8	9	10	11	12	13	14	15	16	17
9	9	10	11	12	13	14	15	16	17	18

$13 - 5 = 8$

★ **ANSWER** Luis has 8 cookies left.

Multiplication and Division Concepts

Wow, our tower has...
Let's see, 3 + 3 + 3 + 3 + 3 + 3.
That's...

Multiplication can be a shortcut for addition. It's a shortcut you can use when all the addends are the same.

The tower has 18 blocks because 6 times 3 is 18.

Not for long!

When you have equal groups, you can **multiply** to find how many in all. This is called **multiplication**.

Here are some reasons you might use multiplication.

- There are equal groups to join.
  Multiply to find how many.

3 nests. **2** eggs in each.
Multiply **2** by **3**.
There are **6** eggs in all.

- There are things in equal rows.
  Multiply to find how many.

More
Help
See
98

2 rows. **5** caps in each row.
Multiply **5** by **2**.
There are **10** caps.

Multiplication Words and Symbols

The numbers to **multiply** are **factors.**

The total number is the **product.**

The **multiplication sign** tells you to multiply.

Think 3 groups. 2 in each group.

Write   factors   product

$$3 \times 2 = 6 \quad \text{or} \quad \begin{array}{r} 2 \\ \times\ 3 \\ \hline 6 \end{array}$$

multiplication sign   factors ... product

Say   *Three times two equals six.*

MATH ALERT

To Multiply You Need Equal Groups

More Help
See 96

Sometimes you add and sometimes you multiply to find a total. Be careful! You can't multiply unless you have equal groups.

- These groups are **equal.**

You can **add.**	You can **multiply.**
3 + 3 + 3 + 3 + 3 = 15	5 × 3 = 15

- These groups are **not equal.**

You must **add.**
3 + 1 + 5 + 2 + 4 = 15

You cannot multiply here.

Learning to Multiply

More Help
See 93

There are many ways to find a product.

Use Models to Multiply

Counters help you see the multiplication.

EXAMPLE 1 $2 \times 3 = \blacksquare$

Think 2×3 means 2 groups of 3.

Count to find how many in all. There are 6 in all.

Say *Two times three equals six.*

★ **ANSWER** $2 \times 3 = 6$

EXAMPLE 2 $5 \times 2 = \blacksquare$

Think 5×2 means 5 groups of 2.

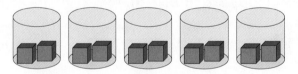

Count. There are 10 in all.

Say *Five times two equals ten.*

★ **ANSWER** $5 \times 2 = 10$

Draw a Picture to Multiply

You can draw a picture to find a product.

More Help
See 93

EXAMPLE $3 \times 5 = $ ■

 One Way Start with the number of groups.

① Draw to show the number of groups.	② Draw to show the number in each group.	③ Count to find the total.
○ ○ ○ 3 groups	5 in each group	15 in all

Another Way Start with one group.

① Draw to show the number in one group.	② Draw the rest of the groups.	③ Count to find the total.
1 group of 5	2 more groups of 5	15 in all

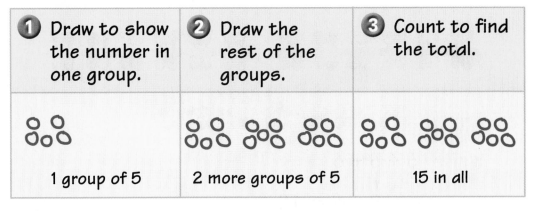

★ **ANSWER** $3 \times 5 = 15$

More Help
See 93

Repeated Addition

You can add the numbers in equal groups to find the product.

Multiplication is a shortcut for adding equal groups.

EXAMPLE 1 $4 \times 5 = \blacksquare$

More Help
See 62, 67, 70

Write $5 + 5 + 5 + 5$

$10 + 10 = 20$

Use strategies to add. Think about doubles.

★ **ANSWER** $4 \times 5 = 20$

EXAMPLE 2 $3 \times 6 = \blacksquare$

Write $6 + 6 + 6$

$12 + 6 = 18$

★ **ANSWER** $3 \times 6 = 18$

Skip Count to Multiply

You can skip count to find a product.

EXAMPLE 1 $4 \times 2 = \blacksquare$

Think 4×2 is 4 twos. Count by twos.

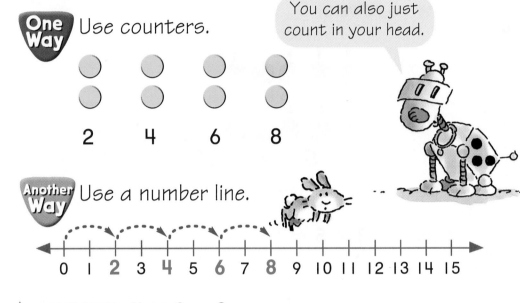

One Way Use counters.

> You can also just count in your head.

2   4   6   8

Another Way Use a number line.

0 1 2 3 4 5 6 7 8 9 10 11 12 13 14 15

★ **ANSWER** $4 \times 2 = 8$

EXAMPLE 2 $3 \times 5 = \blacksquare$

Think 3×5 is 3 fives. Count by fives.

Use groups of 5 counters or use nickels.

More Help See 166

5   10   15

★ **ANSWER** $3 \times 5 = 15$

Arrays

You can make an **array** to find a product.

EXAMPLE 1 $3 \times 5 = \blacksquare$

> An array has equal rows.

Think 3×5 means 3 groups of 5.

Show 3 rows of 5

More Help
See 10, 12

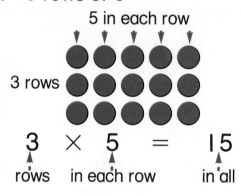

5 in each row

3 rows

> Sometimes this is called a 3 by 5 array.

$$3 \times 5 = 15$$

rows   in each row   in all

★ **ANSWER** $3 \times 5 = 15$

EXAMPLE 2 $2 \times 4 = \blacksquare$

Think 2×4 means 2 groups of 4.

Show 2 rows of 4

4 in each row

2 rows

$$2 \times 4 = 8$$

rows   in each row   in all

★ **ANSWER** $2 \times 4 = 8$

You can color on grid paper to make arrays.

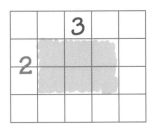

1 row
3 in the row
3 in all
$1 \times 3 = 3$

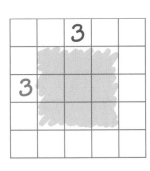

2 rows
3 in each row
6 in all
$2 \times 3 = 6$

3 rows
3 in each row
9 in all
$3 \times 3 = 9$

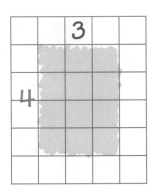

4 rows
3 in each row
12 in all
$4 \times 3 = 12$

This is a huge array we're marching in.

Hooray for our array!

It's also a huge band we're in.

Make a Table to Multiply

You can make a table to find a product.

EXAMPLE 1 How many sock puppets can you make with 4 pairs of socks?

More
Help
See
97

> Remember that there are 2 socks in a pair. So, you can count by twos.

1 pair of socks makes 2 puppets.

Pairs of socks	1	2	3	4
Puppets		2	4	6	8

> The table helps you keep track as you skip count.

1 pair makes 2 puppets.
2 pairs make 4 puppets.
3 pairs make 6 puppets.
4 pairs make 8 puppets.

★ **ANSWER** You can make 8 puppets with 4 pairs of socks.

Did You Know?

Shari Lewis made the first Lamb Chop puppet from an old sock. Lamb Chop later became a big television and video star.

Source: cnn.com/showbiz

EXAMPLE 2 How many finger puppets do you need to fill 4 hands?

 Make a table.

Hands	1	2	3	4
Puppets	5	10	15	20

For this row you can count by fives.

More Help
See 257

 Make an in/out table.

This is also called a **function** table.

Rule: × 5	
In	Out
1	5
2	10
3	15
4	20

In is the number of hands. Out is the number of puppets.

To find the out number, use the rule. Multiply the in number by 5.

★ **ANSWER** You need 20 finger puppets to fill 4 hands.

Multiplication Strategies

More Help
See 93, 263

There are many **strategies**, or methods, you can use to find a product.

Turn-Around Facts in Multiplication

You can multiply in any order.
The product is the same.

Both children are correct.

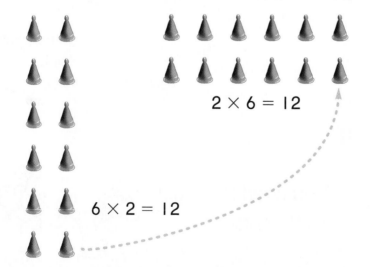

$2 \times 6 = 12$

$6 \times 2 = 12$

The total number of counters is the same.

If you know a multiplication fact, you also know the turn-around fact.

More
Help
See
93

If you know	Then you also know
3 × 2 = 6	2 × 3 = 6
5 × 4 = 20	4 × 5 = 20

Doubles do not have turn-around facts.

2 × 2 = 4   3 × 3 = 9   4 × 4 = 16

The array for every double is a square.

I have 12 pennies.

So do I.

Multiply by 1

More Help
See 263

- When you multiply 1 by any number, the number stays the same.

$$3 \times 1 = \blacksquare$$

Think   3 groups with 1 bird

Write $3 \times 1 = 3$

Say   *Three times one equals three.*

- When you multiply any number by 1, the number stays the same.

$$1 \times 3 = \blacksquare$$

Think   1 group of 3 birds

Write $1 \times 3 = 3$

Say   *One times three equals three.*

I have 5 groups of birds.

Well, I have only 1 group, but I have more birds than you do.

$$5 \times 1 = 5 \qquad 1 \times 6 = 6$$

Multiply by 0

More Help
See 263

- When you multiply 0 by any number, the product is 0.

$$3 \times 0 = \blacksquare$$

Think 3 groups with 0 birds

Write $3 \times 0 = 0$

Say *Three times zero equals zero.*

- When you multiply any number by 0, the product is 0.

$$0 \times 3 = \blacksquare$$

Think 0 groups of 3 birds

Write $0 \times 3 = 0$

Say *Zero times three equals zero.*

You can multiply 0 by any number and get 0.
$99 \times 0 = 0$
$999 \times 0 = 0$
$9999 \times 0 = 0$
A lot of nothing is still nothing!

Use a Multiplication Table

You can find products on a multiplication table.
This table also shows you a picture of each fact.

More
Help
See
93

The second factor is the column number.

The first factor is the row number.

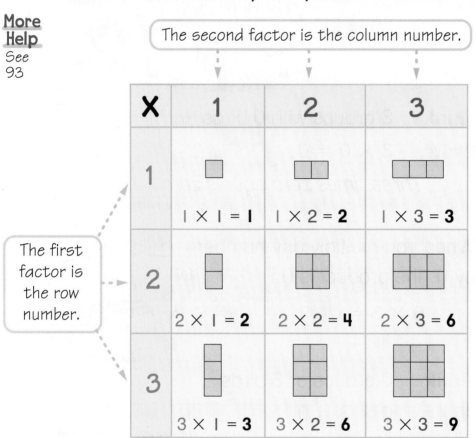

X	1	2	3
1	1 × 1 = 1	1 × 2 = 2	1 × 3 = 3
2	2 × 1 = 2	2 × 2 = 4	2 × 3 = 6
3	3 × 1 = 3	3 × 2 = 6	3 × 3 = 9

EXAMPLE $4 \times 6 = \blacksquare$

1 Find the row for 4.	**2** Find the column for 6. ┄┄┄┄┄	**3** The product is where the row and column meet.

×	1	2	3	4	5	6	7
1	1	2	3	4	5	6	7
2	2	4	6	8	10	12	14
3	3	6	9	12	15	18	21
4	4	8	12	16	20	(24)	28
5	5	10	15	20	25	30	35

★ **ANSWER** $4 \times 6 = 24$

We're waiting for a table.

A multiplication table?

CAFÉ ARRAY
PLEASE wait to be seated.

When you have a group of things, you can **divide** it into smaller groups. This is called **division**.

Here are some reasons you might use division.

- You can divide to find the number in each group.

6 toys.
2 children will share equally.

Divide **6** by **2**.
Each child gets **3** toys.

- You can divide to find the number of groups.

 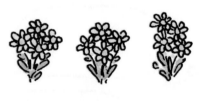

12 flowers.
Put **4** flowers in each bunch.

Divide **12** by **4**.
There are **3** bunches of **4** flowers.

How Many in Each Group?

You can use counters or draw a picture to find how many in each group.

EXAMPLE Four children share 8 balls. How many balls does each child get?

 Act out the problem with counters.

One for you, one for you, one for you, and one for me.

Another one for you, and you, and you, and another one for me. We each get two.

Keep sharing until you run out of counters.

Another Way Draw a picture.

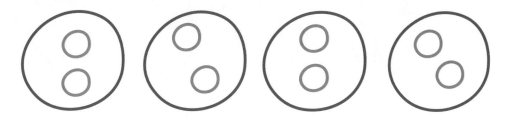

★ **ANSWER** Each child gets 2 balls.

How Many Groups?

You can use counters or draw a picture to find how many groups.

EXAMPLE You have 12 balloons. How many groups of 3 can you make?

 Use counters.

Show 12 counters. Put them into groups of 3.

 Draw a picture.

Draw 12 balloons.

Draw the strings to make groups of 3.

★ **ANSWER** You can make 4 groups of 3 balloons.

Division Sentences

EXAMPLE There are 15 counters. Divide them into groups of 3. How many groups do you make?

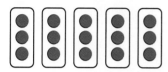

You can write division two ways.

Write $15 \div 3 = 5$ or $3\overline{)15}$ with 5 above

Say   *Fifteen divided by three equals five.*

★ **ANSWER** You make 5 groups of 3.

 **MATH ALERT**

Make Sure You Divide Into Equal Groups

The number in each group you make must be the same.

- Divide 8 counters into 4 groups.

Each group has 2 counters.

Sometimes you have extras.

- Divide 9 counters into 4 groups.

Each group has 2 counters.

There is 1 counter left over.

The extra is the **remainder**.

Addition and Subtraction
with Greater Numbers

You can do exercises like 49 + 1 and 80 + 10 in your head. For harder exercises, some people might use a calculator. But when a calculator isn't handy, it's nice to have a way to add and subtract any numbers.

There are many ways to add and subtract in your head. Try these **mental math** strategies. Then use the strategies that work for you.

Mental Addition

You know how to add numbers like 2 + 2 in your head. You can also learn to add larger numbers mentally.

Add Tens

EXAMPLE 40 + 20 = ■

One Way Think of addition facts.

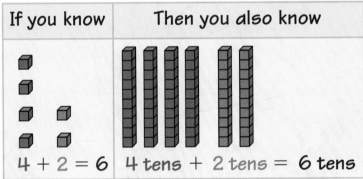

If you know	Then you also know
4 + 2 = 6	4 tens + 2 tens = 6 tens

6 tens is 60.

More Help
See 56

Another Way Use a hundred chart.
Start at 40.
Go *down* 2 rows to *add* 2 tens.

More Help
See
8, 61

1	2	3	4	5	6	7	8	9	10
11	12	13	14	15	16	17	18	19	20
21	22	23	24	25	26	27	28	29	30
31	32	33	34	35	36	37	38	39	40
41	42	43	44	45	46	47	48	49	50
51	52	53	54	55	56	57	58	59	60
61	62	63	64	65	66	67	68	69	70
71	72	73	74	75	76	77	78	79	80

+ 1 ten

+ 2 tens

When you move down 1 row, you add 1 ten, or 10.
When you move down 2 rows, you add 2 tens, or 20.
You can use arrows to show this.
40 ↓ means 40 + 10, or 50.
40 ↓ ↓ means 40 + 20, or 60.

Another Way You might not need to use a chart.
Just count on by tens.
Start at 40. Count on 2 tens.
40 50, 60

More Help
See
97

★ **ANSWER** 40 + 20 = 60

Add Tens to Tens and Ones

EXAMPLE $14 + 30 = \blacksquare$

  Count on by tens.

More Help

See 60–61

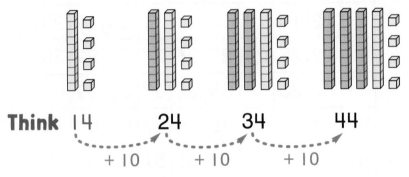

Think 14   24   34   44

$+ 10$ $+ 10$ $+ 10$

 Count on a hundred chart.

11	12	13	14
21	22	23	24
31	32	33	34
41	42	43	44

$+ 1$ ten

$+ 2$ tens

$+ 3$ tens

 Break up the addends.

Think 14 is 1 ten 4 ones.

    30 is 3 tens.

Add the tens. $10 + 30 = 40$

Add the ones to the tens. $40 + 4 = 44$

★ **ANSWER** $14 + 30 = 44$

Add Ones to Tens and Ones

To add small numbers, you can count on by ones.

More Help
See 60–61

EXAMPLE Beth had 38 bugs. She finds 3 more. Now how many bugs does she have?

Think $38 + 3 = \blacksquare$

 Count on by ones.

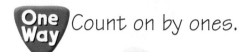

Think 38   39   40   41

+1   +1   +1

Another Way Use a hundred chart.

Start at 38.

Move *across* 3 blocks to *add* 3 ones.

31	32	33	34	35	36	37	38	39	40
41	42	43	44	45	46	47	48	49	50

When you go from 40 to 41, you have to move to the next row. You are still moving across but you've started a new ten.

When you move across 3 blocks you add 3. You can use arrows to show this.
$38 \rightarrow \rightarrow \rightarrow$ means $38 + 3$, or 41.

★ **ANSWER** Now Beth has 41 bugs.

Add Hundreds

You can use addition facts to add hundreds.

EXAMPLE 200 + 300 = ■

If you know	Then you also know
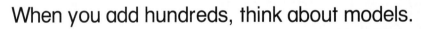 2 + 3 = 5	2 hundreds + 3 hundreds = 5 hundreds

★ **ANSWER** 200 + 300 = 500

More Help
See 20–21, 116

Add Hundreds to Hundreds, Tens, and Ones

When you add hundreds, think about models.

EXAMPLE 1 132 + 200 = ■

Count on from 132 by hundreds.

The number of tens and ones does not change when you add hundreds.

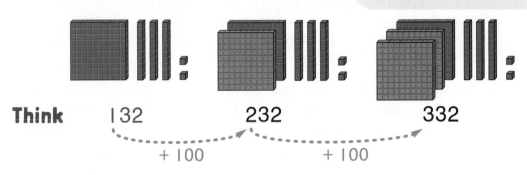

Think   132   232   332

+ 100   + 100

★ **ANSWER** 132 + 200 = 332

EXAMPLE 2 $400 + 539 = $ ■

More
Help
See
58

$400 + 539$ and $539 + 400$
have the same sum.

Count on from 539 by hundreds.

Think 539   639   739   839   939

     + 100   + 100   + 100   + 100

★ **ANSWER** $400 + 539 = 939$

EXAMPLE 3 $300 + 46 = $ ■

More
Help
See
21

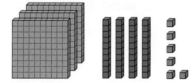

Think $300 + 40 + 6 = 346$

Sometimes you can just "read"
the sum from the addends:
$300 + 46$
three hundred forty-six

★ **ANSWER** $300 + 46 = 346$

$29 + 400 = $ ■.
Hmm, that's
$400 + 29.$ Oh, 429.

???

Why are you
looking at my
head?

I'm trying
to read your
mind!

Mental Subtraction

You can subtract 4 − 2 in your head. You can also subtract larger numbers mentally.

Subtract Tens

EXAMPLE 50 − 20 = ■

 Think of subtraction facts.

More Help
See 74

If you know	Then you also know
5 − 2 = 3	5 tens − 2 tens = 3 tens

 Use a hundred chart.
Start at 50.
Go straight *up* 2 rows to *subtract* 2 tens.

More Help
See 8, 61

21	22	23	24	25	26	27	28	29	30	−2 tens
31	32	33	34	35	36	37	38	39	40	−1 ten
41	42	43	44	45	46	47	48	49	50	

★ **ANSWER** 50 − 20 = 30

Subtract Tens from Tens and Ones

EXAMPLE 47 − 30 = ◼

 Count back by tens.

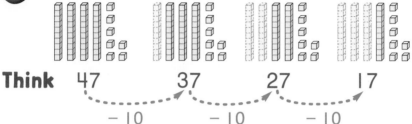

Think 47   37   27   17

− 10   − 10   − 10

Another Way Use a hundred chart.

Start at 47.

Go straight *up* 3 rows to *subtract* 3 tens.

More Help
See 76

11	12	13	14	15	16	17
21	22	23	24	25	26	27
31	32	33	34	35	36	37
41	42	43	44	45	46	47

−3 tens

−2 tens

−1 ten

When you move up 1 block, you subtract 1 ten, or 10.
When you move up 3 blocks, you subtract 3 tens, or 30.
You can use arrows to show this.
47 ↑ means 47 − 10, or 37.
47 ↑ ↑ ↑ means 47 − 30, or 17.

★ **ANSWER** 47 − 30 = 17

Subtract Hundreds

Use subtraction facts to subtract hundreds.

EXAMPLE 1 $400 - 300 = \blacksquare$

If you know	Then you also know
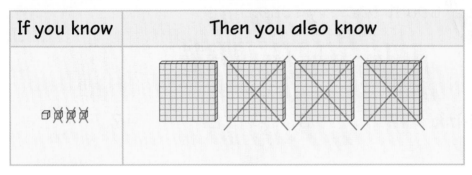	

★ **ANSWER** $400 - 300 = 100$

EXAMPLE 2 How many more cans of nuts did Lake School sell than Valley School?

Cans of Nuts Sold

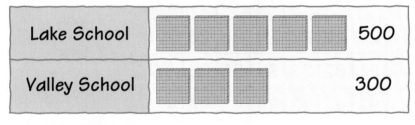

Remember that you can subtract to compare.

Write $500 - 300 = \blacksquare$

<u>More Help</u>
See 80–81

Think $5 - 3 = 2$

5 hundreds − 3 hundreds = 2 hundreds

$500 - 300 = 200$

★ **ANSWER** Lake School sold 200 more cans.

Subtract Hundreds from Hundreds, Tens, and Ones

EXAMPLE 1 432 − 200 = ■

You can count back by hundreds.

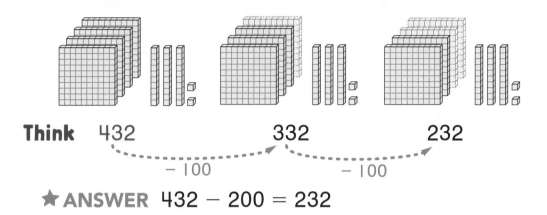

Think   432   332   232

− 100   − 100

★ **ANSWER** 432 − 200 = 232

EXAMPLE 2 Students at Estabrook School sold greeting cards. How many more boxes did Grade 1 sell than Grade 2?

Estabrook School Card Sales

Grade	Number of Boxes
Grade 1	548
Grade 2	200

Write 548 − 200 = ■

You can count back by hundreds.

548 448, 348   So, 548 − 200 = 348

★ **ANSWER** Grade 1 sold 348 more boxes.

You can follow a step-by-step plan to add numbers that have two or more digits.

Add Tens and Ones—No Regrouping

Use what you know about place value and addition facts to add two-digit numbers.

EXAMPLE Amber made 24 little muffins and 12 big muffins for the bake sale. How many muffins did she make in all?

To solve this problem, add 24 and 12.

More Help
See 12–13

24 is 2 tens 4 ones.

Tens	Ones
▭▭	⬜⬜ ⬜⬜

12 is 1 ten 2 ones.

Tens	Ones
▭	⬜⬜

Use models to find the sum.

① Show each addend.	② Join the ones.	③ Join the tens.
Tens / **Ones**	**Tens** / **Ones**	**Tens** / **Ones**
	6 ones	3 tens

3 tens 6 ones = 36

 Add the numbers one place at a time. Begin with the ones.

① Line up the digits.	② Add the ones.	③ Add the tens.
Tens | Ones	Tens | Ones	Tens | Ones
2 | 4	2 | 4	2 | 4
+1 | 2	+1 | 2	+1 | 2
	6	3 | 6
	6 ones	3 tens

★ **ANSWER** Amber made 36 muffins.

 **Line Up the Ones to Add**

```
  3 4
+   5
-----
  3 9
```

 If you write 5 in the tens column, you will be adding 50, not 5!

More Help See 15

Regroup Ones as Tens

More Help
See 137

Sometimes when you add ones you get 10 or more.

Regroup 10 ones as 1 ten.

> Regrouping is sometimes called **trading** or **renaming**.

EXAMPLE 1 You have 15 ones. Regroup so you have fewer than 10 ones.

15 ones

Tens	Ones
	🁢🁢🁢🁢🁢
	🁢🁢🁢🁢🁢
	🁢🁢🁢🁢🁢

Regroup 10 ones as 1 ten.

Tens	Ones
▭	🁢🁢🁢🁢🁢

1 ten 5 ones

★ **ANSWER** 1 ten 5 ones

EXAMPLE 2 You have 1 ten 18 ones. Regroup so you have fewer than 10 ones.

1 ten 18 ones

Tens	Ones
	🁢🁢🁢
	🁢🁢🁢🁢🁢
	🁢🁢🁢🁢🁢
▭	🁢🁢🁢🁢🁢

Regroup 10 ones as 1 ten.

Tens	Ones
	🁢🁢🁢
▭	🁢🁢🁢🁢🁢
▭	🁢🁢🁢
	🁢🁢🁢🁢🁢

2 tens 8 ones

★ **ANSWER** 2 tens 8 ones

Add Tens and Ones—Regroup Ones

When you have 10 or more ones, regroup 10 ones as 1 ten.

More Help
See 139

EXAMPLE 1

$$\begin{array}{r} 27 \\ + 18 \\ \hline \end{array}$$

27 is 2 tens 7 ones.

18 is 1 ten 8 ones.

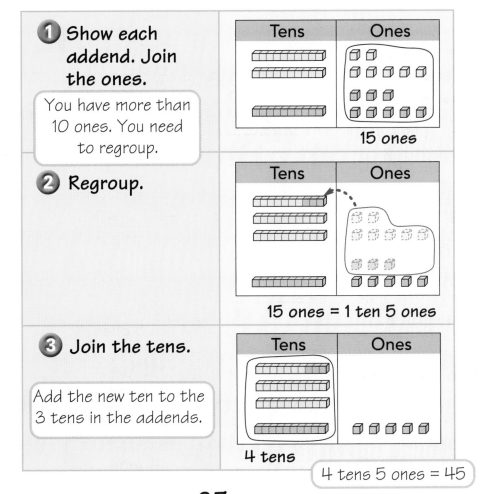

1 Show each addend. Join the ones.

You have more than 10 ones. You need to regroup.

Tens	Ones
	15 ones

2 Regroup.

Tens	Ones

15 ones = 1 ten 5 ones

3 Join the tens.

Add the new ten to the 3 tens in the addends.

Tens	Ones

4 tens

4 tens 5 ones = 45

★ ANSWER

$$\begin{array}{r} 27 \\ + 18 \\ \hline 45 \end{array}$$

MORE ▶

MORE ON **Add Tens and Ones— Regroup Ones**

EXAMPLE 2 Kelly used 16 beads to make a bracelet. She used 27 beads to make a necklace. How many beads did she use?

To solve this problem, add 16 and 27.

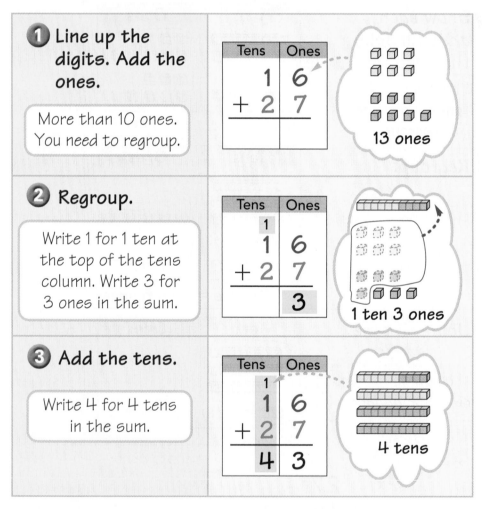

1 Line up the digits. Add the ones.

More than 10 ones. You need to regroup.

Tens	Ones
1	6
+ 2	7

13 ones

2 Regroup.

Write 1 for 1 ten at the top of the tens column. Write 3 for 3 ones in the sum.

Tens	Ones
1
1	6
+ 2	7
	3

1 ten 3 ones

3 Add the tens.

Write 4 for 4 tens in the sum.

Tens	Ones
1
1	6
+ 2	7
4	3

4 tens

★ **ANSWER** Kelly used 43 beads.

EXAMPLE 3 89 + 9 = ■

① Line up the digits. Add the ones.	② Regroup.	③ Add the tens.

Tens	Ones
8	9
+	9

18 ones

	Tens	Ones
	1	
	8	9
+		9
		8

1 ten 8 ones

	Tens	Ones
	1	
	8	9
+		9
	9	8

9 tens

More than 10 ones.
You need to regroup.

★ **ANSWER** 89 + 9 = 98

Did You Know?

You can invent your own rules for adding two-digit numbers. Just be sure that your rule always works.

More Help
See
114–116

Try this method for adding 37 + 28.
Break up the addends.

- Add the tens.   30 + 20 = 50

- Add the ones.   7 + 8 = 15

- Add the two sums.   50 + 15 = 65

So, 37 + 28 = 65.

Add Hundreds, Tens, and Ones— No Regrouping

EXAMPLE 243 + 105 = ■

One Way Use models to find the sum.

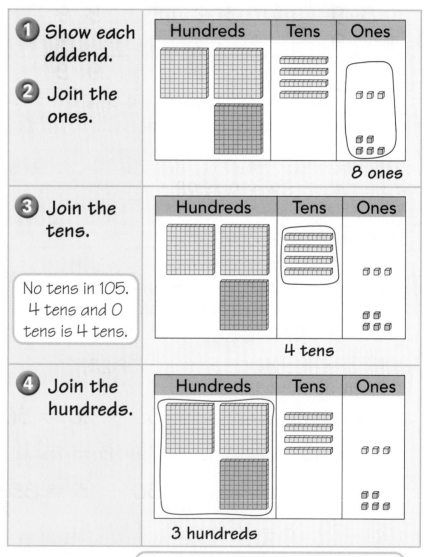

1 Show each addend.

2 Join the ones.

Hundreds	Tens	Ones

8 ones

3 Join the tens.

No tens in 105. 4 tens and 0 tens is 4 tens.

Hundreds	Tens	Ones

4 tens

4 Join the hundreds.

Hundreds	Tens	Ones

3 hundreds

3 hundreds 4 tens 8 ones = 348

Another Way Add the numbers one place at a time.
Begin with the ones.

1 Line up the digits.

Line up the ones.
Line up the tens.
Line up the hundreds.

Hundreds	Tens	Ones
2	4	3
+ 1	0	5

2 Add the ones.

Hundreds	Tens	Ones
2	4	3
+ 1	0	5
		8

8 ones

3 Add the tens.

Hundreds	Tens	Ones
2	4	3
+ 1	0	5
	4	8

4 tens + 0 tens is still 4 tens.

4 tens

4 Add the hundreds.

Hundreds	Tens	Ones
2	4	3
+ 1	0	5
3	4	8

3 hundreds

Ahah!

★ **ANSWER** 243 + 105 = 348

Add Hundreds, Tens, and Ones— Regroup Ones

More Help
See 126, 139

Sometimes the sum of the ones is 10 or more. Regroup 10 ones as 1 ten.

EXAMPLE Raz and his classmates counted the birds that visited the class bird feeder on Monday and Tuesday. How many birds did they count?

Number of Birds Visiting Feeder

Monday	108
Tuesday	374

Add 108 and 374.

1 Add the ones.

Hundreds	Tens	Ones
	1	
1	0	8
+ 3	7	4
		2

12 ones

2 Add the tens.

Hundreds	Tens	Ones
	1	
1	0	8
+ 3	7	4
	8	2

8 tens

3 Add the hundreds.

Hundreds	Tens	Ones
	1	
1	0	8
+ 3	7	4
4	8	2

4 hundreds

You have more than 10 ones. Regroup.
12 ones = 1 ten 2 ones

⭐ **ANSWER** They counted 482 birds.

Regroup Tens as Hundreds

When you have 10 or more *ones*, you can **regroup** 10 *ones* as 1 *ten*.

When you have 10 or more *tens*, you can **regroup** 10 *tens* as 1 *hundred*.

More Help See 126

EXAMPLE You have 1 hundred 13 tens 4 ones. Regroup so you have fewer than 10 tens.

Hundreds	Tens	Ones

1 hundred 13 tens 4 ones

Regroup 10 tens as 1 hundred.

Now you have 2 hundreds.

Hundreds	Tens	Ones

2 hundreds 3 tens 4 ones

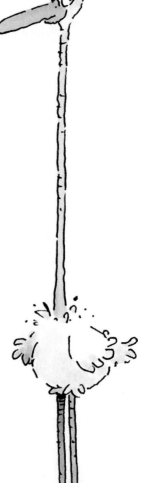

★ **ANSWER** 2 hundreds 3 tens 4 ones

Add—Regroup Tens

When the sum of the tens is 10 or more, you need to regroup tens as hundreds.

EXAMPLE 1 Bridge School PTA had a spaghetti supper. They sold tickets for 275 children and 164 adults. How many tickets did they sell?

More Help See 127, 133

To solve the problem, add 275 and 164.

275 is 2 hundreds 7 tens 5 ones.

164 is 1 hundred 6 tens 4 ones.

	Hundreds	Tens	Ones
1 Show each addend.			

	Hundreds	Tens	Ones
2 Join the ones.			9 ones

3 Join the tens.

More than 10 tens. You need to regroup.

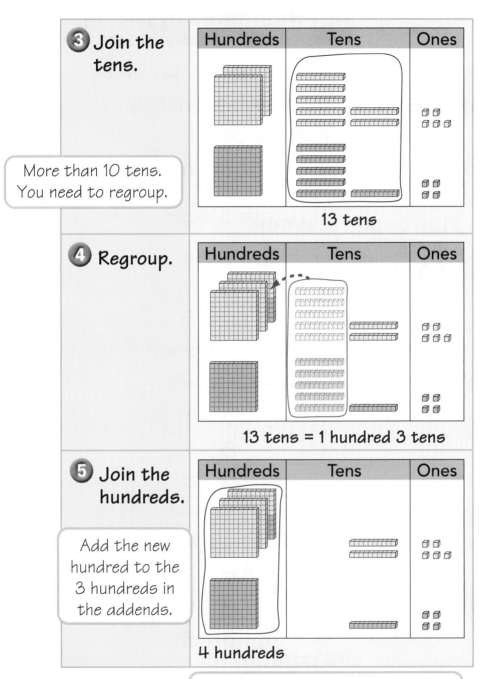

Hundreds	Tens	Ones

13 tens

4 Regroup.

Hundreds	Tens	Ones

13 tens = 1 hundred 3 tens

5 Join the hundreds.

Add the new hundred to the 3 hundreds in the addends.

Hundreds	Tens	Ones

4 hundreds

4 hundreds 3 tens 9 ones = 439

★ ANSWER The PTA sold 439 tickets.

MORE ▶

MORE ON Add—Regroup Tens

EXAMPLE 2 Chen helps at the library. He sorted 571 books last week. This week he sorted 387 books. How many books did he sort?

To solve this problem, add 571 and 387.

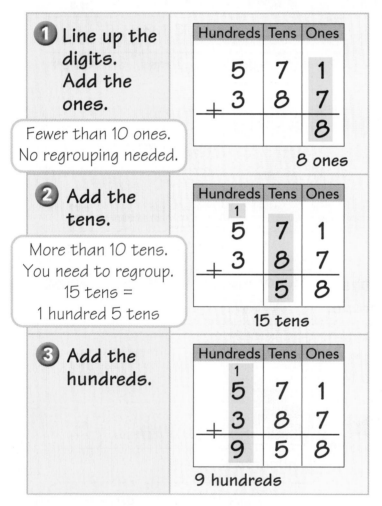

1 Line up the digits. Add the ones.

Fewer than 10 ones. No regrouping needed.

Hundreds	Tens	Ones
5	7	1
+ 3	8	7
		8

8 ones

2 Add the tens.

More than 10 tens. You need to regroup. 15 tens = 1 hundred 5 tens

Hundreds	Tens	Ones
1		
5	7	1
+ 3	8	7
	5	8

15 tens

3 Add the hundreds.

Hundreds	Tens	Ones
1		
5	7	1
+ 3	8	7
9	5	8

9 hundreds

★ **ANSWER** Chen sorted 958 books.

EXAMPLE 3 $84 + 153 = \blacksquare$

Hundreds	Tens	Ones
1		
	8	4
+ 1	5	3
2	3	7

> It's OK to have the lesser number on top.

★ **ANSWER** $84 + 153 = 237$

Sometimes you get a three-digit sum when the addends have only two digits!

$$
\begin{array}{r}
95 \\
+ \ 31 \\
\hline
126
\end{array}
$$

> 9 tens + 3 tens = 12 tens
> Regroup.
> 12 tens = 1 hundred 2 tens

> There are no hundreds to add. You can just write the 1 in the hundreds place in the sum.

Did You Know?

Regrouping in addition used to be called carrying. You carry a digit to the next column.

Today is Monday, May 22, 1956.

$$
\begin{array}{r}
\overset{1}{3}96 \\
+241 \\
\hline
37
\end{array}
$$

> 9 + 4 = 13. Put down the 3 and carry the 1.

Add—Regroup Twice

EXAMPLE 1 75 + 78 = ■

More
Help

See
126,
133,
137

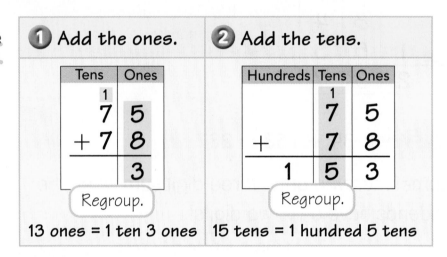

① **Add the ones.**

Tens	Ones
1	
7	5
+ 7	8
	3

Regroup.

13 ones = 1 ten 3 ones

② **Add the tens.**

Hundreds	Tens	Ones
	1	
	7	5
+	7	8
1	5	3

Regroup.

15 tens = 1 hundred 5 tens

★ **ANSWER** 75 + 78 = 153

EXAMPLE 2 109 + 596 = ■

Don't let zeros in a
number trick you!

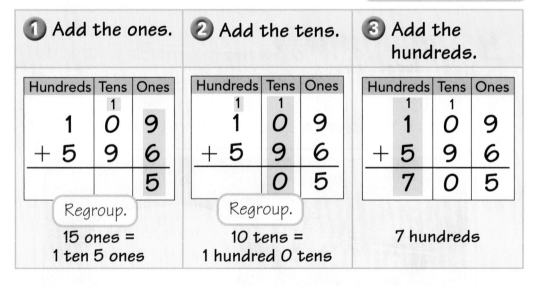

① **Add the ones.**

Hundreds	Tens	Ones
	1	
1	0	9
+ 5	9	6
		5

Regroup.

15 ones =
1 ten 5 ones

② **Add the tens.**

Hundreds	Tens	Ones
1	1	
1	0	9
+ 5	9	6
	0	5

Regroup.

10 tens =
1 hundred 0 tens

③ **Add the hundreds.**

Hundreds	Tens	Ones
1	1	
1	0	9
+ 5	9	6
7	0	5

7 hundreds

★ **ANSWER** 109 + 596 = 705

When You Add, Regroup Only When Necessary

Regroup only when the sum of the digits in a column is 10 or more.

Sometimes you need to regroup and sometimes you don't.

Here you regroup once.	Here you regroup twice.	Here you don't regroup at all.

$$\begin{array}{r} \overset{1}{1}83 \\ +\ 245 \\ \hline 428 \end{array}$$

$$\begin{array}{r} \overset{1\ 1}{3}65 \\ +\ 238 \\ \hline 603 \end{array}$$

$$\begin{array}{r} 406 \\ +\ 152 \\ \hline 558 \end{array}$$

12 tens

13 ones
10 tens

Check Addition

It helps to check your sum to see if it is correct.

EXAMPLE Is Ethan's sum correct?

Ethan

$$549$$
$$+ 201$$
$$\overline{850}$$

One Way Change the order.

More Help See 58

$$\overset{1}{201}$$
$$+ 549$$
$$\overline{750}$$

$1 + 9 = 10$ ones → Regroup.
$1 + 4 = 5$ tens → Do not regroup.
$2 + 5 = 7$ hundreds

Another Way Break up the addends.

More Help See 116

$$549$$
$$+ 201$$
$$\overline{750}$$

Add the hundreds. $500 + 200 = 700$
Add the tens and ones. $49 + 1 = 50$
Add the two sums. $700 + 50 = 750$

★ **ANSWER** Ethan's sum of 850 is *not* correct. It should be 750.

Can you see the error in the addition? 5 tens were regrouped as 1 hundred 5 tens.

Column Addition

More Help
See 70, 262

Remember that you can add in any order.
Look for doubles or other easy facts.

EXAMPLE $38 + 4 + 42 = \blacksquare$

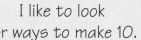

I like to look
for ways to make 10.

1 Line up the digits.	**2** Add the ones.	**3** Add the tens.
$\begin{array}{r} 38 \\ 4 \\ + 42 \\ \hline \end{array}$	$\begin{array}{r} \overset{1}{} \\ 38 \\ 4 \\ + 42 \\ \hline 4 \end{array}$	$\begin{array}{r} \overset{1}{} \\ 38 \\ 4 \\ + 42 \\ \hline 84 \end{array}$

Look for tens.
$8 + 2 = 10$
$10 + 4 = 14$ ones
Regroup.

Think.
$1 + 3 = 4$
$4 + 4 = 8$

★ **ANSWER** $38 + 4 + 42 = 84$

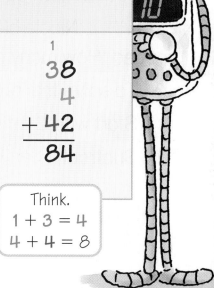

MATH ALERT When You Add, Be Careful When You Regroup

There can be more than 1 ten when you regroup ones.

$\begin{array}{r} \overset{2}{} \\ 29 \\ 18 \\ + 37 \\ \hline 84 \end{array}$

Add the ones.
$9 + 8 + 7 = 24$

Regroup 24 ones
as **2 tens 4 ones.**

Subtraction with Two-Digit and Three-Digit Numbers

You can follow a step-by-step plan to subtract numbers that have two or more digits.

Subtract Tens and Ones— No Regrouping

Use what you know about place value and subtraction facts to subtract two-digit numbers.

EXAMPLE Nicole blew up 27 balloons. 14 balloons popped. How many balloons does she have left?

To solve this problem, subtract 14 from 27.

Start with 2 tens 7 ones.

Subtract 1 ten 4 ones.

One Way 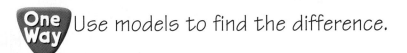 Use models to find the difference.

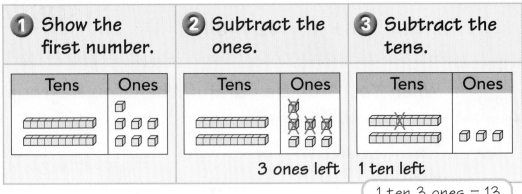

① Show the first number.	② Subtract the ones.	③ Subtract the tens.
Tens / Ones	Tens / Ones	Tens / Ones
	3 ones left	1 ten left

1 ten 3 ones = 13

Another Way Subtract the numbers one place at a time. Begin with the ones.

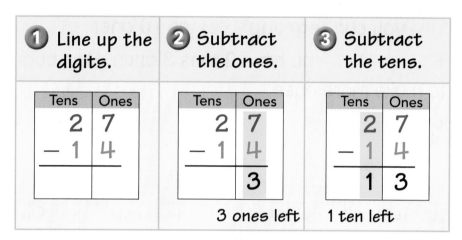

① Line up the digits.	② Subtract the ones.	③ Subtract the tens.

Tens	Ones
2	7
− 1	4

Tens	Ones
2	7
− 1	4
	3

3 ones left

Tens	Ones
2	7
− 1	4
1	3

1 ten left

★ **ANSWER** Nicole has 13 balloons left.

MATH ALERT

Line Up the Ones to Subtract

$48 − 5 = \blacksquare$

You can use squared paper to help you.

$$\begin{array}{r} 4\ 8 \\ -\quad 5 \\ \hline 4\ 3 \end{array}$$

4 tens 8 ones

no tens 5 ones

If you write 5 in the *tens* column, you will have 48 − 50!

More Help See 15

Regroup Tens as Ones

When there are not enough ones to subtract, you can **regroup** I ten as I0 ones.

EXAMPLE 1 You have 2 tens 3 ones. Regroup I ten to make more ones.

More Help
See 126

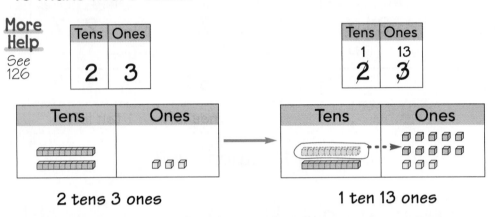

2 tens 3 ones   1 ten 13 ones

★ **ANSWER** I ten I3 ones

EXAMPLE 2 You have 5 tens 0 ones. Regroup I ten to make more ones.

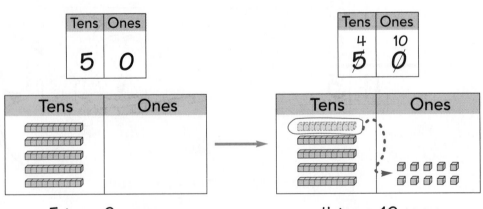

5 tens 0 ones   4 tens 10 ones

★ **ANSWER** 4 tens I0 ones

Subtract Tens and Ones— Regroup Tens

Sometimes you need to subtract more ones than you have. Regroup 1 ten as 10 ones.

EXAMPLE 1 35 − 17 = ■

Start with 3 tens 5 ones.
Subtract 1 ten 7 ones.

1 Not enough ones to subtract 7. Regroup 1 ten.	Tens | Ones 3 tens 5 ones = 2 tens 15 ones
2 Subtract the ones.	Tens | Ones 8 ones left
3 Subtract the tens.	Tens | Ones 1 ten left

1 ten 8 ones = 18

★ **ANSWER** 35 − 17 = 18

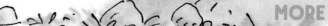

MORE ▶

MORE ON **Subtract Tens and Ones— Regroup Tens**

EXAMPLE 2 34 − 16 = ■

1 Line up the digits. Not enough ones to subtract 6. Regroup.	Tens | Ones 2 | 14 3 | 4 − 1 | 6	 2 tens 14 ones
2 Subtract the ones.	Tens | Ones 2 | 14 3 | 4 − 1 | 6 | 8	8 ones left
3 Subtract the tens.	Tens | Ones 2 | 14 3 | 4 − 1 | 6 1 | 8	 1 ten left

★ **ANSWER** 34 − 16 = 18

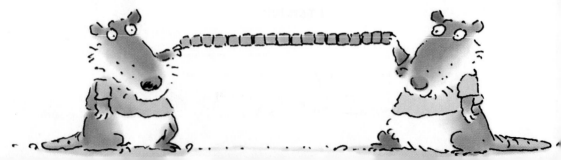

EXAMPLE 3 David Shannon wrote a book that was printed in two languages. Find the missing number in this chart.

More Help See 74, 84–85

Half Moon Library

Title	Checked Out
Duck on a Bike English Version © 2002, New York.	34 times
Pato Va en Bici Spanish Version © 2002, Editorial Juventod, Barcelona.	■ times
Total for both versions	62 times

Think $34 + ■ = 62$ or

$62 - 34 = ■$

Subtract to find the missing part.

1 Not enough ones to subtract 4. Regroup.

Tens	Ones
5	12
6̸	2
– 3	4

5 tens 12 ones

2 Subtract the ones.

Tens	Ones
5	12
6̸	2
– 3	4
	8

8 ones

3 Subtract the tens.

Tens	Ones
5	12
6̸	2
– 3	4
2	8

2 tens

★ **ANSWER** The missing number is 28.

Subtract Hundreds, Tens, and Ones—No Regrouping

EXAMPLE 1 428 − 103 = ■

More Help
See 142

1 Show the first number.

2 Subtract the ones.

103 has 3 ones.

Hundreds	Tens	Ones

5 ones left

3 Subtract the tens.

103 has 0 tens. No tens to subtract.

Hundreds	Tens	Ones

2 tens

4 Subtract the hundreds.

103 has 1 hundred.

Hundreds	Tens	Ones

3 hundreds left

3 hundreds 2 tens 5 ones = 325

★ **ANSWER** 428 − 103 = 325

EXAMPLE 2 A second-grade class made 297 cards to sell. They sold 65 cards the first day. How many cards did they have left?

Subtract. 297 − 65 = ■

1 Line up the digits. **2** Subtract the ones.	Hundreds	Tens	Ones
	2	9	7
	−	6	5
	2	3	2

2 ones left

3 Subtract the tens.	Hundreds	Tens	Ones
	2	9	7
	−	6	5
	2	3	2

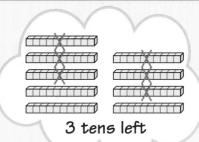

3 tens left

4 Subtract the hundreds.	Hundreds	Tens	Ones
No hundreds to subtract. 2 − 0 = 2	2	9	7
	−	6	5
	2	3	2

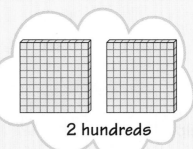

2 hundreds

⭐ **ANSWER** They had 232 cards left.

Subtracting is faster than counting all the cards that are left!

More Help
See 144, 155

Subtract Hundreds, Tens, and Ones—Regroup Tens

Sometimes you need to subtract more ones than you have. Regroup 1 ten as 10 ones.

EXAMPLE 1 481 − 157 = ■

① Not enough ones. Regroup 1 ten. Subtract the ones.	② Subtract the tens.	③ Subtract the hundreds.

Hundreds	Tens	Ones
	7	11
4	8̸	1̸
− 1	5	7
		4

4 ones

Hundreds	Tens	Ones
	7	11
4	8̸	1̸
− 1	5	7
	2	4

2 tens

Hundreds	Tens	Ones
	7	11
4	8̸	1̸
− 1	5	7
3	2	4

3 hundreds

★ **ANSWER** 481 − 157 = 324

EXAMPLE 2 540 − 231 = ■

More Help
See 22

Hundreds	Tens	Ones
	3	10
5	4̸	0̸
− 2	3	1
3	0	9

Remember that zeros are important.

★ **ANSWER** 540 − 231 = 309

Regroup Hundreds as Tens

When you need more *ones*, you can **regroup**
1 *ten* as 10 *ones*.

When you need more *tens*, you can **regroup**
1 *hundred* as 10 *tens*.

More
Help
See
144

EXAMPLE You have 2 hundreds 6 tens 4 ones.
Regroup 1 hundred to make more tens.

Hundreds	Tens	Ones
2	6	4

Hundreds	Tens	Ones

2 hundreds 6 tens 4 ones

Hundreds	Tens	Ones
1	16	
2	6	4

Hundreds	Tens	Ones

1 hundred 16 tens 4 ones

★ **ANSWER** 1 hundred 16 tens 4 ones

More
Help
See
145,
151

Subtract—Regroup Hundreds

Sometimes you need to regroup 1 hundred as 10 tens.

EXAMPLE 428 − 253 = ■

1 Show the first number. **2** Subtract the ones.	Hundreds	Tens		Ones
			5 ones left	

3 Not enough tens. Regroup 1 hundred.	Hundreds	Tens		Ones
	10 tens and 2 tens = 12 tens			

4 Subtract the tens. **5** Subtract the hundreds.	Hundreds	Tens		Ones
	1 hundred left	7 tens left		

1 hundred 7 tens 5 ones = 175

★ **ANSWER** 428 − 253 = 175

EXAMPLE 2 Hua Mei, a giant panda, weighed 73 pounds on her first birthday. Her mother Bai Yun weighed 249 pounds on that day. How much less did Hua Mei weigh than her mother did?

Subtract to compare.

$249 - 73 = \blacksquare$ 236

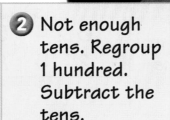

	① Line up the digits. Subtract the ones.	② Not enough tens. Regroup 1 hundred. Subtract the tens.	③ Subtract the hundreds.

You can just bring down the 1 hundred.

Hundreds	Tens	Ones
2	4	9
−	7	3
		6

6 ones

Hundreds	Tens	Ones
¹	¹⁴	
2	4	9
−	7	3
	7	6

7 tens

Hundreds	Tens	Ones
	1	14
2	4	9
−	7	3
1	7	6

1 hundred

★ **ANSWER** Hua Mei weighed 176 pounds less than her mother did.

More Help

See 218–219

Did You Know?

Hua Mei weighed about 5 ounces when she was born at the San Diego Zoo.

Source: sandiegozoo.org/special/pandas

Subtract—Regroup Twice

Sometimes you have to regroup tens and hundreds in the same exercise.

EXAMPLE 645 − 248 = ■

1 Not enough ones to subtract 8. Regroup 1 ten. Subtract the ones.

Hundreds	Tens	Ones
6	³4̷	¹⁵5̷
− 2	4	8
		7

2 Not enough tens to subtract 4. Regroup 1 hundred. Subtract the tens.

Hundreds	Tens	Ones
⁵6̷	¹³3̷	¹⁵5̷
− 2	4	8
	9	7

3 Subtract the hundreds.

Hundreds	Tens	Ones
⁵6̷	¹³3̷	¹⁵5̷
− 2	4	8
3	9	7

★ **ANSWER** 645 − 248 = 397

When You Subtract, Regroup Only When Necessary

Sometimes you need to regroup and sometimes you don't.

> When you subtract, you must look carefully at the two digits in *each column*. If the digit on the bottom is greater than the digit on the top, you must regroup.

7 13
$$
\begin{array}{r}
3\cancel{8}\cancel{3} \\
- 245 \\
\hline
138
\end{array}
$$

Here you regroup once.

5 is greater than 3.

13
8 $\cancel{9}$ 15
$$
\begin{array}{r}
9\cancel{4}\cancel{5} \\
- 687 \\
\hline
258
\end{array}
$$

Here you regroup twice.

7 is greater than 5.

8 is greater than 3.

$$
\begin{array}{r}
456 \\
- 152 \\
\hline
304
\end{array}
$$

Here you don't regroup at all.

Subtract Across Zeros

Sometimes you need to regroup tens, but you find there are no tens! First you have to regroup hundreds. Then you can regroup tens.

EXAMPLE 1 $403 - 186 = $ ■

1 Not enough ones to subtract 6. You need to regroup 1 ten as 10 ones. There are no tens! Regroup 1 hundred to make 10 tens.	<table><tr><th>Hundreds</th><th>Tens</th><th>Ones</th></tr><tr><td>3 4̶</td><td>10 0̶</td><td>3</td></tr><tr><td>− 1</td><td>8</td><td>6</td></tr><tr><td></td><td></td><td></td></tr></table>
2 Now regroup 1 ten as 10 ones.	<table><tr><th>Hundreds</th><th>Tens</th><th>Ones</th></tr><tr><td>3 4̶</td><td>9 10̶ 0̶</td><td>13 3</td></tr><tr><td>− 1</td><td>8</td><td>6</td></tr><tr><td>3</td><td>8</td><td>9</td></tr></table>
3 Subtract ones, tens, and hundreds.	<table><tr><th>Hundreds</th><th>Tens</th><th>Ones</th></tr><tr><td>3 4̶</td><td>9 10̶ 0̶</td><td>13 3</td></tr><tr><td>− 1</td><td>8</td><td>6</td></tr><tr><td>2</td><td>1</td><td>7</td></tr></table>

★ **ANSWER** $403 - 186 = 217$

EXAMPLE 2 600 − 359 = ■

Hundreds	Tens	Ones
5	9 10̶	10
6̶	0̸	0̸
− 3	5	9
2	4	1

Regroup 1 hundred as 10 tens.
Regroup 1 ten as 10 ones.

★ **ANSWER** 600 − 359 = 241

I regrouped 60 tens as 59 tens 10 ones.

$$\begin{array}{r} {}^{59}\ {}^{10} \\ 6\cancel{0}\cancel{0} \\ -\ 359 \\ \hline 241 \end{array}$$

Check Subtraction

You can use addition to check your subtraction.

EXAMPLE Is Lee's difference correct?

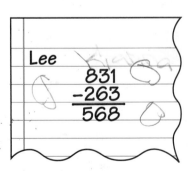

Lee
831
−263
568

Add the number you subtracted to the difference. You should get the number you started with.

$$\begin{array}{r} 831 \\ -\ 263 \\ \hline 568 \end{array} \quad\times\quad \begin{array}{r} 568 \\ +\ 263 \\ \hline 831 \end{array}$$

★ **ANSWER** Yes, Lee's difference of 568 is correct.

Estimate when you want to know *about* how many.

Estimate Sums

You can use what you know about rounding numbers to estimate sums.

EXAMPLE 1 There were 33 first graders on the playground. There were 48 second graders on the playground. About how many children were on the playground?

Use rounding to estimate the sum for 33 + 48.

More Help
See 40

Round each addend to the nearest 10.

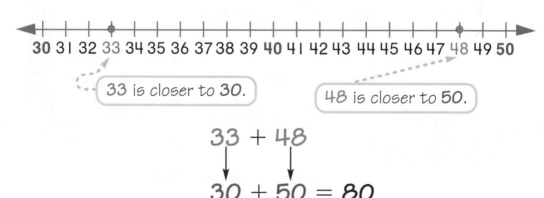

You can use a number line to round.

30 31 32 33 34 35 36 37 38 39 40 41 42 43 44 45 46 47 48 49 50

33 is closer to 30.

48 is closer to 50.

$$33 + 48$$
$$30 + 50 = 80$$

★ **ANSWER** There were about 80 children on the playground.

You can estimate a sum in different ways. You may not always get the same **estimate**.

More Help
See 41

EXAMPLE 2 Estimate the sum for 428 + 495.

 One Way Round each addend to the nearest hundred.

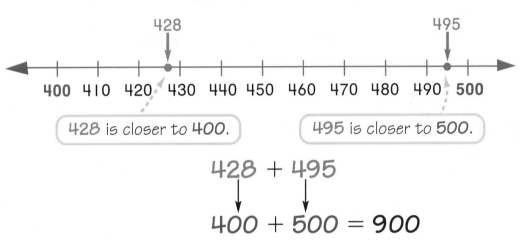

428

495

400 410 420 430 440 450 460 470 480 490 500

428 is closer to 400.

495 is closer to 500.

428 + 495

400 + 500 = 900

Another Way Use front digits to estimate.

428 + 495

Keep the first, or front, digit in each addend. Make the other digits zeros.

400 + 400 = 800

Some people call this **front-end** estimation.

Actual sum: 923
Rounding may give you a closer estimate. But front-end estimation can be faster!

★ **ANSWER** The sum is about 800 or 900.

Estimate Differences

You can use rounding to estimate differences.

EXAMPLE 1 Chad had 57 stamps. He used 28 of them. About how many stamps does he have left?

More Help
See 40 Use rounding to estimate the difference for 57 − 28.

Use a number line to round each number to the nearest 10.

40 41 42 43 44 45 46 47 48 49 **50** 51 52 53 54 55 56 57 58 59 **60**

57 is closer to **60**.

20 21 22 23 24 25 26 27 28 29 **30** 31 32 33 34 35 36 37 38 39 **40**

28 is closer to **30**.

$$57 - 28$$
$$\downarrow \qquad \downarrow$$
$$60 - 30 = 30$$

★ **ANSWER** Chad has about 30 stamps left.

You can estimate a difference in more than one way. You may not always get the same estimate.

EXAMPLE 2 Estimate the difference for 787 − 331.

More Help
See 41

One Way Round each number to the nearest hundred.

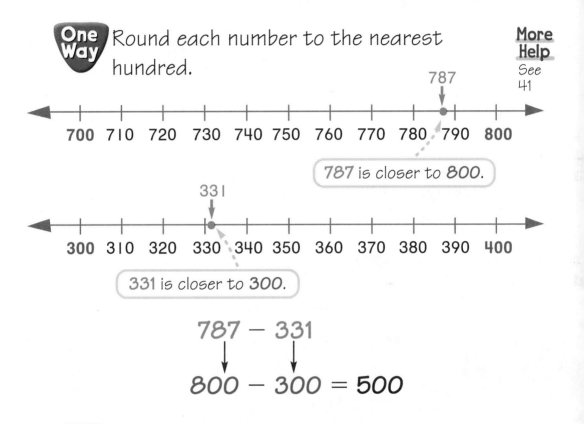

787 is closer to 800.

331 is closer to 300.

$$787 - 331$$
$$800 - 300 = 500$$

Another Way Use front digits to estimate.

$$787 - 331$$
$$700 - 300 = 400$$

Keep the first, or front, digit in each number. Make the other digits zero.

★ **ANSWER** The difference is about 500 or 400.

Money and Time

Money ————————— 164

Time ————————— 180

What time is it?

It's 3 o'clock!

We use numbers with money.
A toy costs 2 dollars. You only
have 75 cents.

We also use numbers with time.
A movie starts at 5 o'clock. It takes
20 minutes to get to the theater.

Sometimes we use numbers with
money and time together,
even when we wish we didn't.

That will be
10 dollars for
the clock, please.

"IF YOU BREAK IT
YOU BUY IT!"

You use **money** to buy or pay for things.

Values of Coins and Bills

United States coins and bills have values that are given in **cents** and in **dollars**.

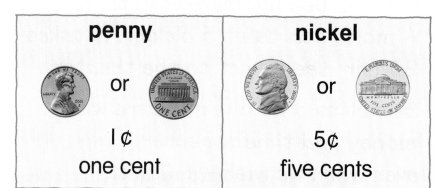

penny	nickel
or	or
1¢	5¢
one cent	five cents

> 1 nickel has the same value as 5 pennies.

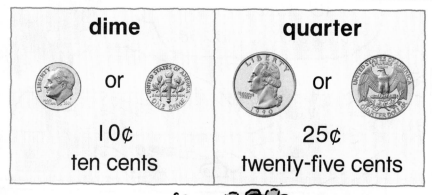

dime	quarter
or	or
10¢	25¢
ten cents	twenty-five cents

> A dime may be small, but it is worth more than a penny or a nickel.

half-dollar

or

50¢
fifty cents

dollar

or

$1.00
one dollar

> 1 dollar is worth 100¢. That's the same value as 100 pennies.

Writing Dollars and Cents

Write 28¢ or $0.28

cent sign   dollar sign   decimal point

Say *twenty-eight cents*

> When you use a cent sign, don't use a dollar sign or a decimal point.

Write $1.35   (You say *and* for the decimal point.)

Say *one dollar and thirty-five cents*

Write $3.04

Say *three dollars and four cents*

Did You Know?

There is also a coin that is worth one dollar.

Count Money

You can **skip count** and **count on** to find the value of a group of coins.

Count Dimes, Nickels, or Pennies

EXAMPLE 1 How much are 6 dimes worth?

10¢ 20¢ 30¢ 40¢ 50¢ 60¢

1 dime = 10¢
So, you can skip count by tens.

★ **ANSWER** 60¢ or $0.60

EXAMPLE 2 How much are 6 nickels worth?

5¢ 10¢ 15¢ 20¢ 25¢ 30¢

1 nickel = 5¢
So, you can skip count by fives.

★ **ANSWER** 30¢ or $0.30

EXAMPLE 3 How much are 6 pennies worth?

★ **ANSWER** 6¢ or $0.06

1 penny = 1¢
So, 6 pennies = 6¢.

Count On with Pennies

EXAMPLE 1 How much is this set of coins worth?

Skip count by fives.   Then count on by ones.

5¢   10¢   15¢   16¢   17¢   18¢

More Help
See 60, 97

★ **ANSWER** 18¢ or $0.18

EXAMPLE 2 How much is this set of coins worth?

Skip count by tens.   Then count on by ones.

10¢   20¢   30¢   40¢   41¢   42¢

★ **ANSWER** 42¢ or $0.42

EXAMPLE 3 How much is this set of coins worth?

Skip count by tens.   Add five.   Then count on by ones.

10¢   20¢   25¢   26¢   27¢

★ **ANSWER** 27¢ or $0.27

Count Quarters

Four quarters
are worth one dollar.
A quarter is
one fourth of a dollar.

More Help
See 43, 46–47

Two quarters equal
half of a dollar.

EXAMPLE How much are 6 quarters worth?

> You can skip count by 25s.

More Help
See 165

| 25¢ | 50¢ | 75¢ | 100¢ | 125¢ | 150¢ |

100¢ = $1.00   150¢ = $1.50

Remember, you write dollars and cents
with a dollar sign and a decimal point.

> That's one dollar and fifty cents.

★ **ANSWER** 6 quarters are worth $1.50.

Did You Know?

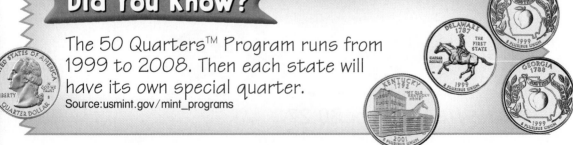

The 50 Quarters™ Program runs from
1999 to 2008. Then each state will
have its own special quarter.
Source: usmint.gov/mint_programs

Count On From a Quarter

EXAMPLE 1 What is 1 quarter and 2 dimes worth?

Count on by tens.

25¢   35¢   45¢

> You can count by tens, fives, or ones from 25¢.

More Help
See 60, 97

★ **ANSWER** 45¢ or $0.45

EXAMPLE 2 What is 1 quarter and 3 nickels worth?

Count on by fives.

25¢   30¢   35¢   40¢

★ **ANSWER** 40¢ or $0.40

EXAMPLE 3 What is 1 quarter and 3 pennies worth?

Count on by ones.

25¢   26¢   27¢   28¢

★ **ANSWER** 28¢ or $0.28

Count Mixed Groups of Coins

When you count coins, it is easier to start with coins that are worth the most.

EXAMPLE 1 What are 2 dimes, 1 quarter, and 2 nickels worth?

More Help
See 166–167, 169

Think Start counting with the quarter.

Count on by tens.   Count on by fives.

25¢   35¢   45¢   50¢   55¢

★ **ANSWER** 55¢ or $0.55

EXAMPLE 2 Carlos found 3 nickels, 1 penny, and 2 dimes in his old jacket pocket. How much money did he find?

Think Start counting with the dimes.

Skip count by tens.   Count on by fives.   Count one more.

10¢   20¢   25¢   30¢   35¢   36¢

★ **ANSWER** He found 36¢ or $0.36.

EXAMPLE 3 How much are these coins worth?

You can put coins in order to make them easier to count.

More Help

See 166–167, 169

Count on by tens.

Count on by fives.

Count on by ones.

25¢ 35¢ 40¢ 45¢ 46¢ 47¢ 48¢ 49¢ 50¢

★ **ANSWER** 50¢ or $0.50

EXAMPLE 4 A pencil costs 26¢. Show a group of coins you can use to pay for the pencil.

One Way Use any coins you want.

There are other groups of coins. You could use 5 nickels and 1 penny or 26 pennies.

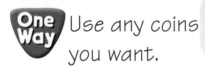

Another Way Use the fewest coins.

Count Bills and Coins

EXAMPLE 1 How much money is in this wallet?

More Help
See 164–169

Think Count the dollars. $1, $2, $3

Count the coins. 25¢ 50¢ 60¢

Write $3.60

Say *three dollars and sixty cents*

★ **ANSWER** There is $3.60 in the wallet.

Remember that you write dollars and cents with a dollar sign and a decimal point.

Did You Know?

There are other United States bills.

EXAMPLE 2 How much money is this?

1 Count the dollar bills.	$1.00 $1.00 $2.00
2 Group coins to make dollars.	25¢ 25¢ 25¢ 25¢ 4 quarters = $1.00 10¢ 10¢ 10¢ 10¢ 10¢ 10¢ 10¢ 10¢ 10¢ 10¢ 10 dimes = $1.00
3 Count the other coins.	1¢ 1¢ 1¢ 3 pennies = $0.03
4 Count to find the total amount.	$2.00 $3.00 $4.00 $4.03 +$1.00 +$1.00 +$0.03

★ **ANSWER** There is $4.03.

More Help

See 127–129

Add Money

You can add money amounts the same way you add two-digit and three-digit numbers.

EXAMPLE 1 What is the total cost of these two toys?

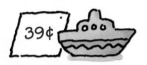 39¢

 45¢

It helps to use dimes and pennies as models.

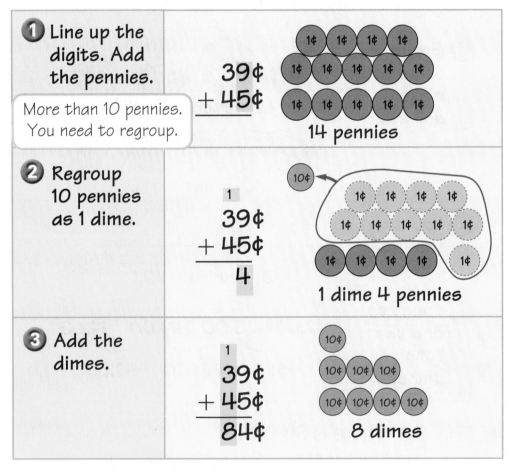

1. **Line up the digits. Add the pennies.**

   More than 10 pennies. You need to regroup.

   $$\begin{array}{r} 39¢ \\ + 45¢ \\ \hline \end{array}$$

   14 pennies

2. **Regroup 10 pennies as 1 dime.**

   $$\begin{array}{r} {\scriptstyle 1} \\ 39¢ \\ + 45¢ \\ \hline 4 \end{array}$$

   1 dime 4 pennies

3. **Add the dimes.**

   $$\begin{array}{r} {\scriptstyle 1} \\ 39¢ \\ + 45¢ \\ \hline 84¢ \end{array}$$

   8 dimes

★ **ANSWER** The total cost is 84¢.

EXAMPLE 2 $\$1.63 + 3.72 = \blacksquare$

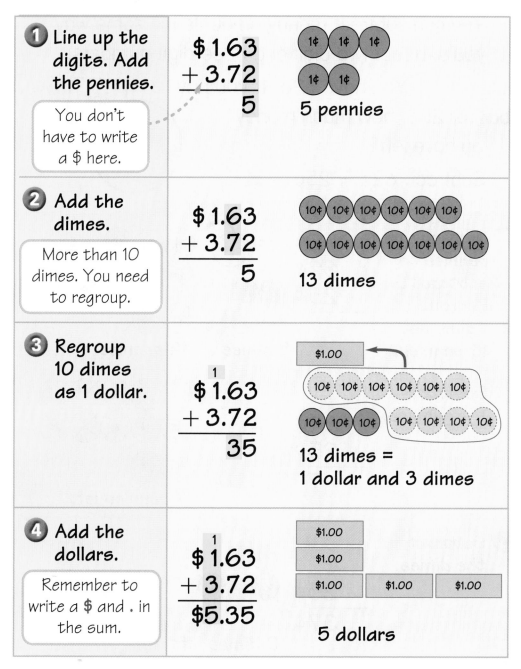

1 Line up the digits. Add the pennies.	$\$1.63$ $+\,3.72$ $\underline{5}$	1¢ 1¢ 1¢ 1¢ 1¢ **5 pennies**
You don't have to write a $ here.		
2 Add the dimes.	$\$1.63$ $+\,3.72$ $\underline{5}$	10¢ 10¢ 10¢ 10¢ 10¢ 10¢ 10¢ 10¢ 10¢ 10¢ 10¢ 10¢ 10¢ **13 dimes**
More than 10 dimes. You need to regroup.		
3 Regroup 10 dimes as 1 dollar.	$\overset{1}{}$ $\$1.63$ $+\,3.72$ $\underline{35}$	$\$1.00$ 10¢ 10¢ 10¢ 10¢ 10¢ 10¢ 10¢ 10¢ 10¢ 10¢ 10¢ 10¢ 10¢ **13 dimes =** **1 dollar and 3 dimes**
4 Add the dollars.	$\overset{1}{}$ $\$1.63$ $+\,3.72$ $\underline{\$5.35}$	$\$1.00$ $\$1.00$ $\$1.00$ $\$1.00$ $\$1.00$ **5 dollars**
Remember to write a $ and . in the sum.		

★ **ANSWER** $\$1.63 + \$3.72 = \$5.35$

More
Help
See
145–
147

Subtract Money

You can subtract money amounts the same way you subtract two-digit and three-digit numbers.

EXAMPLE 1 You have 62¢. If you buy a ball for 25¢, how much money will you have left?

25¢

Subtract. 62¢ − 25¢ = ■

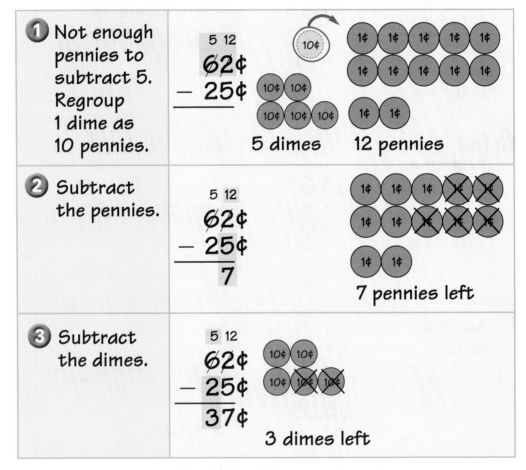

1 Not enough pennies to subtract 5. Regroup 1 dime as 10 pennies.	5 12 6̸2̸¢ − 25¢ 5 dimes 12 pennies
2 Subtract the pennies.	5 12 6̸2̸¢ − 25¢ 7 7 pennies left
3 Subtract the dimes.	5 12 6̸2̸¢ − 25¢ 37¢ 3 dimes left

★ **ANSWER** You will have 37¢ left.

EXAMPLE 2 $4.09 – $1.23 = ■

1 Line up the digits. Subtract the pennies.	$4.09 – 1.23 ⎯⎯ 6	1¢ 1¢ 1¢ ⨂ ⨂ ⨂ 1¢ 1¢ 1¢ ⨂ 6 pennies left
2 Not enough dimes to subtract 2. Regroup 1 dollar as 10 dimes.	3 10 $4.09 – 1.23 ⎯⎯ 6	$1.00 $1.00 $1.00 $1.00 10¢ 10¢ 10¢ 10¢ 10¢ 10¢ 10¢ 10¢ 10¢ 10¢ 3 dollars 10 dimes
3 Subtract the dimes.	3 10 $4.09 – 1.23 ⎯⎯ 86	10¢ 10¢ 10¢ 10¢ ⨂ 10¢ 10¢ 10¢ 10¢ ⨂ 8 dimes left
4 Subtract the dollars.	3 10 $4.09 – 1.23 ⎯⎯ $2.86	$1.00 $1.00 $1.00⨂ 2 dollars left

★ **ANSWER** $4.09 – $1.23 = $2.86

More
Help

*See
152–
153*

Compare Money Amounts

More coins don't always mean more money. One quarter is worth more money than 3 nickels.

EXAMPLE 1 Who has more money? Who has less money?

More Help See 28–29, 166

Think 7¢ < 11¢ and 11¢ > 7¢

| < means "is less than." | > means "is greater than" |

⭐ **ANSWER** Juan has more money than Pat. Pat has less money that Juan.

EXAMPLE 2 Who has more money?

More Help See 28–29, 170

Think 25¢ = 25¢

⭐ **ANSWER** Mio and Lionel have the same amount.

Make Change

You get **change** when you give a store clerk more money than an item costs.

EXAMPLE You buy a 34¢ stamp. You give the clerk 50¢. How much change do you get?

 Start at the price. Count up to the amount you give the clerk.

A penny, a nickel and a dime. That's 16¢

More Help
See 145–147, 176

Another Way Subtract.

$$\begin{array}{r} \overset{4\;10}{\cancel{5}\cancel{0}}¢ \\ -\;34¢ \\ \hline 16¢ \end{array}$$

★ **ANSWER** You get 16¢ in change.

You have a time to be at school, a time for lunch, a time for bed.

Clocks and Time

Clocks measure **time**. A clock face has numbers around the edge and hands that move.

- The **hour hand** is the short hand.

The hour hand shows the hour.

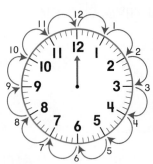

It takes one hour to go from one number to the next.

It takes 12 hours to go around the clock.

- The **minute hand** is the long hand.

The minute hand shows the minutes before and after the hour.

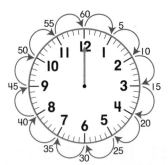

It takes 5 minutes to go from one number to the next.

It takes 60 minutes to go around the clock.

A **digital clock** does not have hands.

It shows you the hours and minutes.

hours ⋯ minutes

A colon separates the hours from the minutes.

Minutes and Hours

 60 minutes = 1 hour   24 hours = 1 day

- ## How long is a **minute**?

1 Mississippi, 2 Mississippi, 3 Mississippi ...

It takes about one minute to count to 60 Mississippi.

- ## How long is an **hour**?

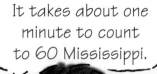

My two favorite TV shows are on for 1 hour.

My favorite TV show is on for 1 hour.

- ## How long is a **day**?

A day is the amount of time between sunrise today and sunrise tomorrow.

Or, the time between lunchtime today and lunchtime tomorrow.

Tell Time

You read a clock to tell what time it is.

Time to the Hour

When the minute hand on a clock face is on the 12, you say the time as "o'clock."

School starts at 8 o'clock.

Reading period starts at 10 o'clock.

Lunch is at 12 o'clock.

Write 12 o'clock or 12:00

Say *twelve o'clock*

If it is daytime, 12 o'clock is also called 12 **noon**, or just noon. If it is nighttime, it's called 12 **midnight**, or just midnight.

Time to the Half Hour

60 minutes = 1 hour   30 minutes = 1 half hour

See

The hour hand points halfway between the 8 and 9.

The minute hand points to the 6. It is halfway around the clock.

The hour is 8. ⟍⟍⟍⟍⟋⟋⟋ The minutes show 30.

This is 30 minutes after the hour.

Write 8:30

Say   *eight thirty*

You can also write and say half past 8 or 30 minutes after 8.

Are we there yet?

We'll be there in half an hour.

You said that 30 minutes ago.

That was when we were moving.

Time to Five Minutes

See

You can skip count by fives to find the time.

Write 4:25

Say *four twenty-five*

You could also say 25 minutes after 4.

More Help
See 186

See

 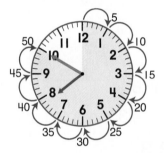

You could also say p

Write 7:50

Say *seven fifty*

These digital clocks also show 4:25 and 7:50.

Did You Know?

All the numbers on a digital clock are made from the same seven dashes.

1 2 3 4 5 6 7 8 9 0

1 uses only 2 dashes.   8 uses all of the dashes.

Time to the Quarter Hour

15 minutes = 1 quarter hour

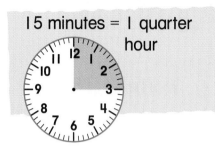

30 minutes = 1 half hour
or
2 quarters
of an hour

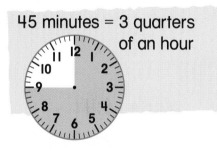

45 minutes = 3 quarters of an hour

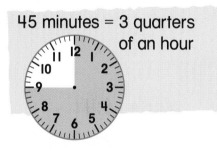

60 minutes = 1 hour

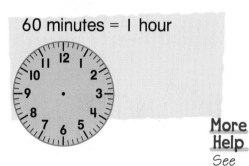

More Help
See 186

Write	7:15	7:30	7:45
Say	*seven fifteen or quarter past seven*	*seven thirty or half past seven*	*seven forty-five or quarter to eight*

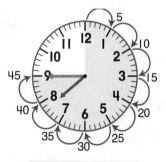

You could also say 15 minutes after 7.

You could also say 30 minutes after 7.

You could also say 15 minutes to 8.

Time to the Minute

60 minutes = 1 hour

See

First skip count by fives as far as you can.

Then count on by ones.

Write 1:18

Say *one eighteen*

See

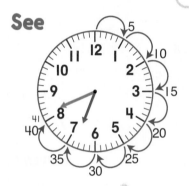

Write 6:41

Say *six forty-one*

MATH ALERT

Read the Hour Correctly

As the time gets closer to the hour, the hour hand gets closer to the next number. The hour hand is closer to the 4 than to the 3. But, the time is 3:45, not 4:45.

Find a Later Time

EXAMPLE 1 It is 3:15. What time will it be in 10 minutes?

 3:15

Count on 10 more minutes.

3:25

⭐ **ANSWER** 3:25

EXAMPLE 2 It is 11:45. What time will it be in a half hour?

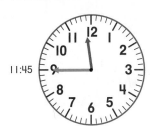 11:45

 12:15

Count on 30 minutes. Or, just move halfway around the clock.

⭐ **ANSWER** 12:15

A.M. and P.M.

Times between midnight and noon are A.M.

10:00 A.M.

Times between noon and midnight are P.M.

10:00 P.M.

Elapsed Time

You can count on from one time to another to see how much time passes. The time that passes is called **elapsed time**.

EXAMPLE 1 How long is Nikki's party?

Please come to
my **PARTY**

When: Saturday

Time: 2 P.M. to 4 P.M.

Place: My House

from Nikki

Nikki's party starts at 2:00.

Nikki's party ends at 4:00.

Count hours.

2:00 to 3:00 is 1 hour.

3:00 to 4:00 is 1 hour.

★ **ANSWER** Nikki's party is 2 hours long.

Schedules may list the starting and ending times of activities. You can find how long each activity lasts by finding the elapsed time.

Soaring Scientists Camp

More
Help
See
187

Activity	Start	End
Rub-a-Dub	9:00 A.M.	10:30 A.M.
Foil Boats	9:30 A.M.	10:30 A.M.
Bubble Shapes	10:30 A.M.	1:00 P.M.
Squawk Band	1:00 P.M.	4:00 P.M.

EXAMPLE 2 How long does Bubble Shapes last?

Bubble Shapes starts
at 10:30 A.M.

Bubble Shapes ends
at 1:00 P.M.

Count hours.

　　10:30 A.M. to 11:30 A.M. is 1 hour.

　　11:30 A.M. to 12:30 P.M. is 1 hour.

Then count minutes.

　　12:30 P.M. to 1:00 P.M. is 30 minutes.

★ **ANSWER** Bubble Shapes lasts
2 hours and 30 minutes.

You could also say
it lasts $2\frac{1}{2}$ hours.

Calendar

A **calendar** is used to keep track of days, weeks, and months of a year.

More Help
See 181

Days of the Week

There are 7 days in a week.

The days are Sunday, Monday, Tuesday, Wednesday, Thursday, Friday, and Saturday.

EXAMPLE What day of the week is January 15?

More Help
See 35

Write January 15

Say *January fifteenth*

① Find 15.

② Go to the top of the column to find the day.

These are short ways to write the days of the week.

JANUARY						
Sun.	Mon.	Tues.	Wed.	Thurs.	Fri.	Sat.
				1	2	3
4	5	6	7	8	9	10
11	12	13	14	15	16	17
18	19	20	21	22	23	24
25	26	27	28	29	30	31

★ **ANSWER** Thursday

Did You Know?

Martin Luther King, Jr. was born on January 15, 1929. His holiday is celebrated on the third Monday of January.
Source: www.holidays.net/mlk/holiday

Months of the Year

There are 12 months in a year.

1 January

S	M	T	W	T	F	S
		1	2	3	4	5
6	7	8	9	10	11	12
13	14	15	16	17	18	19
20	21	22	23	24	25	26
27	28	29	30	31		

2 February

S	M	T	W	T	F	S
					1	2
3	4	5	6	7	8	9
10	11	12	13	14	15	16
17	18	19	20	21	22	23
24	25	26	27	28		

3 March

S	M	T	W	T	F	S
					1	2
3	4	5	6	7	8	9
10	11	12	13	14	15	16
17	18	19	20	21	22	23
24	25	26	27	28	29	30
31						

4 April

S	M	T	W	T	F	S
	1	2	3	4	5	6
7	8	9	10	11	12	13
14	15	16	17	18	19	20
21	22	23	24	25	26	27
28	29	30				

5 May

S	M	T	W	T	F	S
			1	2	3	4
5	6	7	8	9	10	11
12	13	14	15	16	17	18
19	20	21	22	23	24	25
26	27	28	29	30	31	

6 June

S	M	T	W	T	F	S
						1
2	3	4	5	6	7	8
9	10	11	12	13	14	15
16	17	18	19	20	21	22
23	24	25	26	27	28	29
30						

7 July

S	M	T	W	T	F	S
	1	2	3	4	5	6
7	8	9	10	11	12	13
14	15	16	17	18	19	20
21	22	23	24	25	26	27
28	29	30	31			

8 August

S	M	T	W	T	F	S
				1	2	3
4	5	6	7	8	9	10
11	12	13	14	15	16	17
18	19	20	21	22	23	24
25	26	27	28	29	30	31

9 September

S	M	T	W	T	F	S
1	2	3	4	5	6	7
8	9	10	11	12	13	14
15	16	17	18	19	20	21
22	23	24	25	26	27	28
29	30					

10 October

S	M	T	W	T	F	S
		1	2	3	4	5
6	7	8	9	10	11	12
13	14	15	16	17	18	19
20	21	22	23	24	25	26
27	28	29	30	31		

11 November

S	M	T	W	T	F	S
					1	2
3	4	5	6	7	8	9
10	11	12	13	14	15	16
17	18	19	20	21	22	23
24	25	26	27	28	29	30

12 December

S	M	T	W	T	F	S
1	2	3	4	5	6	7
8	9	10	11	12	13	14
15	16	17	18	19	20	21
22	23	24	25	26	27	28
29	30	31				

The calendar shows you the number of days in each month. Usually, February has 28 days; but in a **leap year**, February has 29 days. The years 2008, 2012, 2016, and 2020 are leap years.

Geometry

Maybe square wheels are not such a good idea.

Why are wheels round? Why do bricks and blocks have flat faces? We use these shapes because sometimes rolling is important and sometimes stacking is important. If you don't think so, just imagine using different shapes for some things.

There are different kinds of lines. You stand in line. You write on lines. You might use a computer to go online.

In geometry, a **line** is a straight path that never ends. You can draw a picture of a line.

- A **horizontal line** goes straight across. Think of the lines on your writing paper.

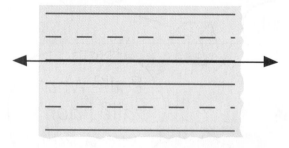

- A **vertical line** goes straight up and down. Think of your writing paper turned sideways.

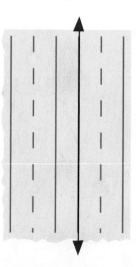

Lines can cross each other.

Some lines will never cross.

Sometimes two lines cross to make square corners. You can use the corners of an index card to check for square corners.

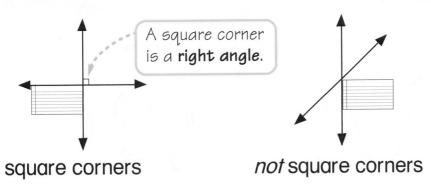

A square corner is a **right angle**.

square corners

not square corners

A **plane figure** is flat, on a flat surface.

Not like an airplane.

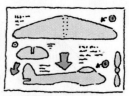

Names of Plane Figures

You probably learned the names of plane figures when you were little. If you look, you will see plane figures everywhere.

Triangle

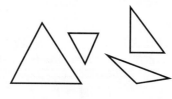

Rectangle

Rectangles have 4 square corners. That's 4 right angles.

More Help
See 195

A **square** is a special kind of rectangle. All the sides are the same length.

Rhombus

All sides are the same length.

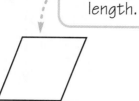

Circle

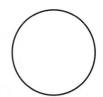

Open and Closed Figures

A plane figure can be an **open figure** or a
closed figure.

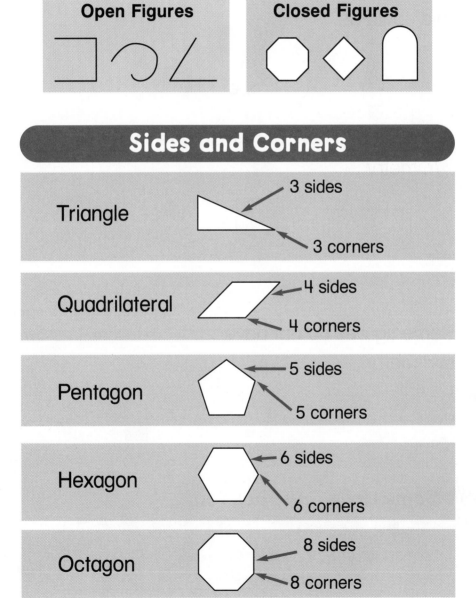

Open Figures

Closed Figures

Sides and Corners

Triangle
3 sides
3 corners

Quadrilateral
4 sides
4 corners

Pentagon
5 sides
5 corners

Hexagon
6 sides
6 corners

Octagon
8 sides
8 corners

Symmetry

If one part of a figure matches the other part exactly when it is folded, the figure has **line symmetry**. The **line of symmetry** is where the fold could be.

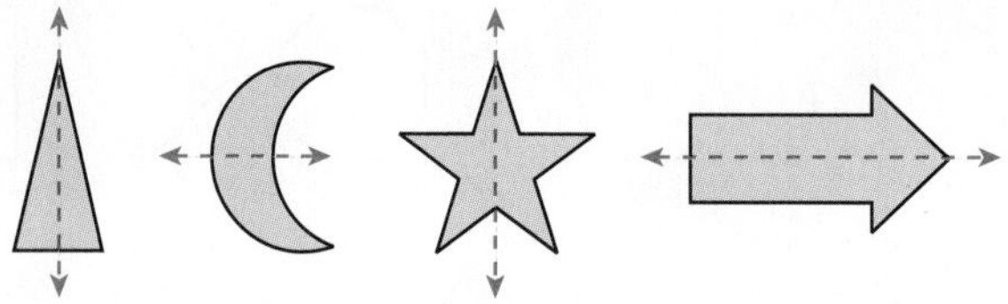

A figure can have more than one line of symmetry.

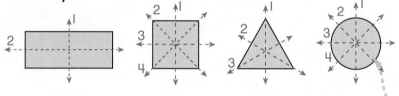

Some figures do not have symmetry.

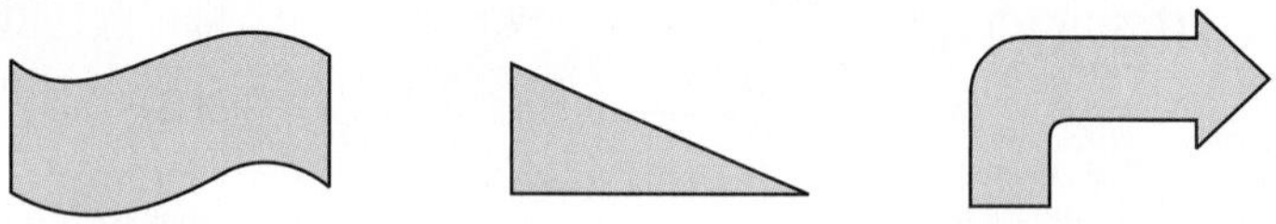

A circle has too many lines of symmetry to count. Here are 4 of them.

More
Help
See
43

Some Lines Can Fool You

This line cuts the rectangle in half. But it is *not* a line of symmetry.

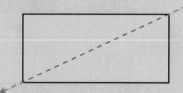

You can make a shape that shows symmetry.

 One Way Fold paper, draw, and cut.

1 Fold a piece of paper.	
2 Draw along the folded edge.	
3 Cut along the line you drew. Don't cut the fold!	
4 Unfold. The line of symmetry is the fold line.	

 Another Way Draw to show the matching part.

Congruent Figures

Figures are **congruent** when they are the same size and same shape.

Congruent	Not Congruent
Same size and same shape	Same size and different shape
Same size and same shape	Different size and same shape
Same size and same shape	Different size and different shape

Sometimes it is hard to see if two figures are congruent. You can trace one figure. Then put the tracing over the other figure. If they are the same size and same shape, they are congruent. Sometimes you need to turn or flip one of the figures.

More Help

See 202–203

You can draw pairs of congruent figures.

EXAMPLE 1 Use tracing paper to draw a pair of congruent figures.

★ ANSWER

Same size.
Same shape.

EXAMPLE 2 Use dot paper to draw a pair of congruent figures.

Use the dots as a guide when you draw.

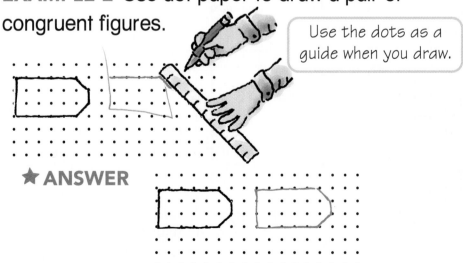

★ ANSWER

Slides, Flips, and Turns

You can move in different ways.

Watch me slide! *Watch me flip.* *Watch me turn.*

A figure can also be moved in different ways.

- You can **slide** a figure.

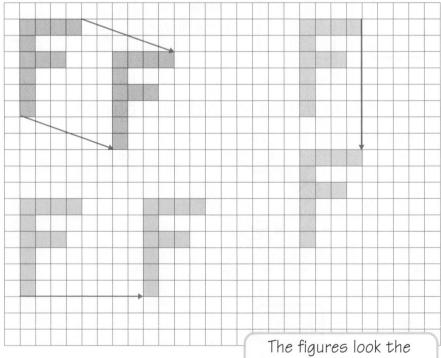

The figures look the same. They just moved.

- You can **flip** a figure across a line.

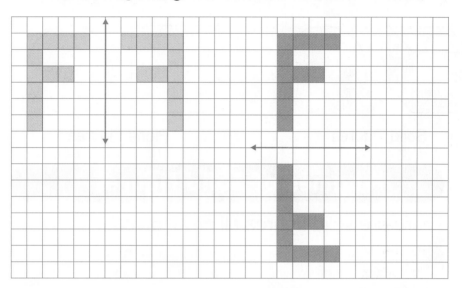

- You can **turn** a figure.

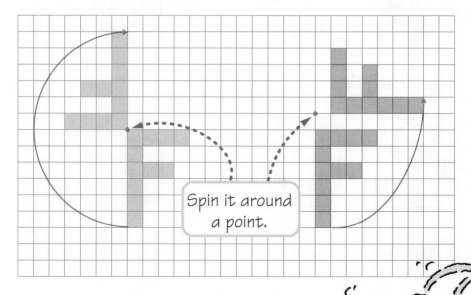

Spin it around
a point.

A **solid figure** is *not* flat like a plane figure.

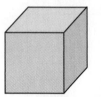  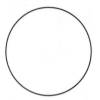

| Solid Figures | Plane Figures |

Names of Solid Figures

There are solid figures all around you.

Spheres

Cubes

Rectangular Prisms

 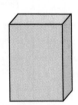

Cylinders

Cones

Pyramids

Faces of Solids

The **faces** of a solid figure are in the shape of plane figures.

More Help
See 196

an edge

a corner

a face

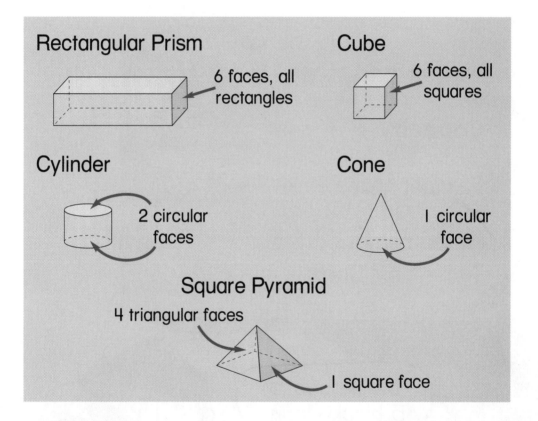

Rectangular Prism

6 faces, all rectangles

Cube

6 faces, all squares

Cylinder

2 circular faces

Cone

1 circular face

Square Pyramid

4 triangular faces

1 square face

Did You Know?

A soccer ball is made of 12 pentagons and 21 hexagons.

Measurement

My mouse is
5 blocks long.

Y ou use measurement to find out things like how long or how tall or how heavy.

Measuring can be useful, but be careful. When you give a measurement, make sure people know what you mean.

- You can compare two things.

The pencil is **longer** than the crayon.

The crayon is **shorter** than the pencil.

- Sometimes you have more than two things.

The brush is the **longest** object in this group.

The chalk is the **shortest** of these objects.

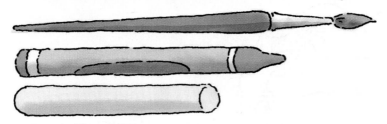

The rabbit is the **shortest** toy animal in this group.

The monkey is **taller** than the rabbit.

The giraffe is the **tallest.**

Measure Length Using Nonstandard Units

You can measure the **length** of an object. Choose a **unit**. Find out how many units long the object is.

EXAMPLE How long is the crayon?

 Use a paper clip as a unit.

Don't overlap units. Put the objects end-to-end when you measure.

Another Way Use a cube as a unit.

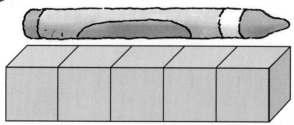

Measures are different when you use units of different sizes, but the object is still the same length.

★ **ANSWER** The crayon is about 3 paper clips long or about 5 cubes long.

Customary Units of Length

You can use rulers with the units **inch**, **foot**, and **yard** to measure length.

Customary Units of Length

1 inch	⊢———————⊣ 1 inch long
1 foot	= 12 inches
1 yard	= 3 feet

Feet means more than one foot.

Estimate and Measure in Inches

Find something that measures about 1 inch. Use it to help you estimate length in inches.

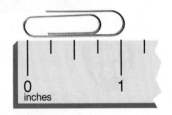

EXAMPLE Estimate the length of the pencil. Then measure with an inch ruler.

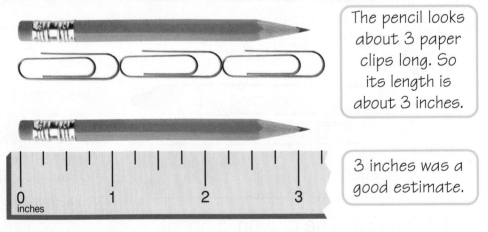

The pencil looks about 3 paper clips long. So its length is about 3 inches.

3 inches was a good estimate.

★ **ANSWER** The pencil is a little more than 3 inches long.

Feet and Yards

Some workbooks are about 1 foot long.

A baseball bat is about 1 yard long.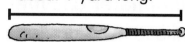

Use small units to measure short things. Use larger units to measure long things.

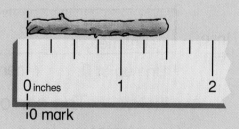

Measure the height of a doll in inches.

Measure the height of a child in feet and inches or just inches.

Measure the length of a path in yards or feet.

MATH ALERT

Line Up Your Ruler Correctly

Measure from the 0 mark.

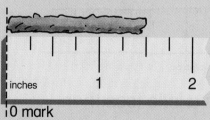

inches   1   2
0 mark

0 inches   1   2
0 mark

Some rulers don't show 0.

0 may not be at the edge of the ruler.

Metric Units of Length

You can use rulers with **centimeter** and **meter** units to measure length.

Metric Units of Length

1 centimeter
1 meter = 100 centimeters

1 centimeter long

Estimate and Measure in Centimeters

Your finger is about 1 centimeter wide.
You can use this to help you estimate length in centimeters.

EXAMPLE Estimate the length of the carrot.
Then measure with a centimeter ruler.

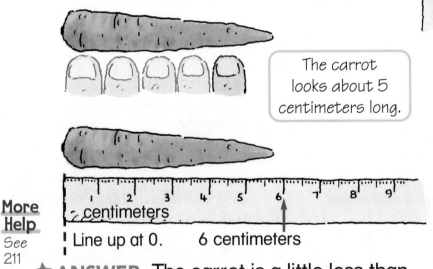

The carrot looks about 5 centimeters long.

centimeters

Line up at 0.   6 centimeters

More Help
See 211

★ **ANSWER** The carrot is a little less than 6 centimeters long.

Meters

A meter is about the distance
from the floor to a doorknob.

Use meters to measure long things. Use
centimeters to measure short things.

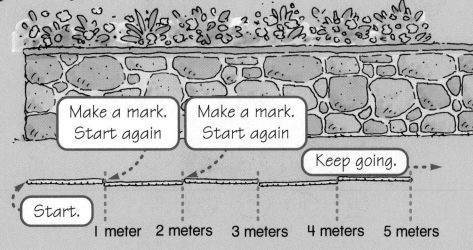

Measure the height of a doll in centimeters.

Measure the height of a child in meters or centimeters.

Measure the length of a path in meters.

MATH ALERT

Keep Track of Your Measures

You can measure something that is
longer than your ruler.

Make a mark. Start again

Make a mark. Start again

Keep going.

Start.

| 1 meter | 2 meters | 3 meters | 4 meters | 5 meters |

You can measure the distance around a figure. You can also measure the amount of space a figure covers.

More Help
See 196

Perimeter

Perimeter is the distance around a figure.

EXAMPLE 1 Bart's garden is in the shape of a triangle. What is the perimeter of his garden?

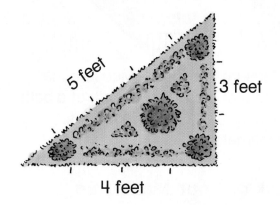

5 feet

3 feet

4 feet

One Way Count the units marked around the figure.

5 feet

3 feet

4 feet

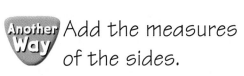 Add the measures of the sides.

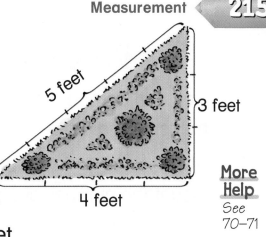

5 feet

3 feet

$5 + 3 + 4 = 12$

4 feet

More Help See 70–71

★ **ANSWER** The perimeter of Bart's garden is 12 feet.

More Help See 196

EXAMPLE 2 The ladybug is walking all the way around the rectangle. How many inches will she walk?

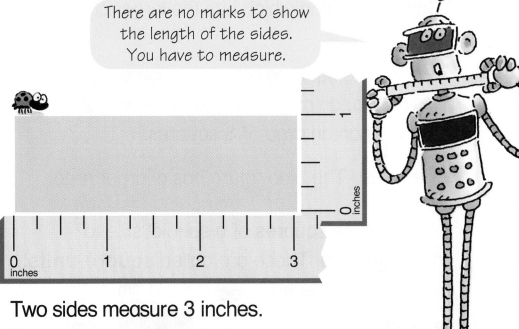

There are no marks to show the length of the sides. You have to measure.

Two sides measure 3 inches.

Two sides measure 1 inch.

Add to find the perimeter.

$3 + 3 + 1 + 1 = 8$

★ **ANSWER** The ladybug will walk 8 inches.

Area

You can compare the size of plane figures by measuring **area.**

EXAMPLE 1 Which figure has a larger area?

Count the number of squares in each figure.

There are 8 squares.   There are 9 squares.

<u>More Help</u>
See 30

Compare to find which has the larger area.

Think   9 > 8

The squares are all the same size. So, an area of 9 squares is larger than an area of 8 squares.

★ **ANSWER** The blue figure has a larger area.

You can use squares of a special size to measure area. These are called **square units.**

EXAMPLE 2 What is the area of the green figure?

★ **ANSWER** The area of the green figure is 4 square centimeters.

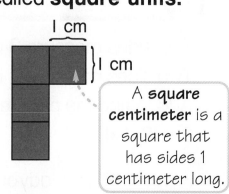

A **square centimeter** is a square that has sides 1 centimeter long.

Be Careful When Comparing Areas

Some areas look larger than others.
But you often have to count square units
to be sure.

More Help
See
30

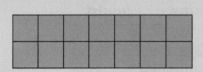

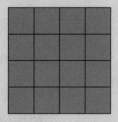

Area = 14 square units   Area = 16 square units.

The orange figure may *look* larger than the
purple figure. The orange figure is longer, but
the purple figure has a larger area because
16 > 14.

Use the same size units when you compare areas.

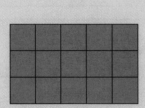

Area = 15 square units   Area = 4 square units

15 > 4, but that doesn't mean the green figure
is larger. The yellow figure has the larger area.

You can measure how heavy something is.

And you can measure how light something is.

Customary Units of Weight

Customary Units of Weight

| ounce |
| 1 pound = 16 ounces |

A loaf of bread weighs about 1 **pound**. So, the box of cookies weighs about 1 pound.

The eraser weighs less than 1 pound.

A slice of bread weighs about 1 **ounce**. So, the pencil weighs about 1 ounce.

The book weighs more than 1 ounce.

Use a **scale** to measure the number of pounds or ounces something weighs.

Weigh small things, like a peach, on a scale that measures in ounces.

Weigh large things, like a person, on a scale that measures in pounds.

You can estimate weight.

EXAMPLE Estimate. Would a bag of six apples weigh about 2 ounces or 2 pounds?

Think   A slice of bread is about 1 ounce.
An apple is heavier than a slice of bread.
So, 6 apples must be heavier than 2 slices of bread. 2 ounces is too light.

★ **ANSWER** 2 pounds

Did You Know?

In 1990, a watermelon that weighed 262 pounds was grown in Tennessee.

Source: watermelon.org

Metric Units of Mass

Metric Units of Mass

gram
1 kilogram = 1000 grams

A math book has a mass of about 1 **kilogram**. So, the apples have a mass of about 1 kilogram.

The banana has a mass that is less than 1 kilogram.

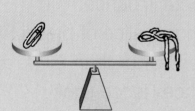

A paper clip has a mass of about 1 **gram**. So, the shoelace has a mass of about 1 gram.

The quarter has a mass that is greater than 1 gram.

Use a **balance** to measure the mass of things in kilograms or grams.

You measure small amounts in grams.

You measure large amounts in kilograms.

You can estimate mass.

EXAMPLE 1 Estimate. Would a dog's mass be closer to 20 grams or 20 kilograms?

> **Think** A dog's mass would be a lot more than the mass of 20 paper clips.

★ **ANSWER** 20 kilograms

EXAMPLE 2 Estimate. Would a tomato have a mass of about 300 grams or 300 kilograms?

★ **ANSWER** 300 grams

The mass of a polar bear is about 300 kilograms!

Capacity is the amount of liquid a container can hold.

Sometimes you compare to find which container holds more. Sometimes you measure the capacity.

Customary Units of Capacity

Customary Units of Capacity

cup
1 pint = 2 cups
1 quart = 2 pints

The milk carton holds I **cup**. So, the mug holds more than I cup.

The apple juice can holds I **pint**. So, the apple juice box holds less than I pint.

The orange juice carton holds I **quart**. So, the water bottle holds more than I quart.

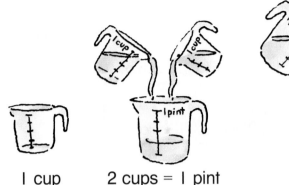

I cup   2 cups = I pint   2 pints = I quart

EXAMPLE 1 Which holds more—two cup measures or I quart measure?

Think   2 cups = I pint   4 cups = I quart

★ **ANSWER** A I quart measure holds more.

EXAMPLE 2 How many cups in 2 quarts?

Think   I quart = 4 cups   2 quarts = 8 cups

★ **ANSWER** 8 cups

More Help
See 43, 46–47

Did You Know?

You should eat 2 to 4 servings of fruit each day. Here are some examples of the size of a serving.

I piece of fruit. $\frac{3}{4}$ cup of juice $\frac{1}{2}$ cup of canned fruit $\frac{1}{4}$ cup of dried fruit

Source: U.S. Department of Agriculture

Metric Units of Capacity

Metric Units of Capacity

milliliter
1 liter = 1000 milliliters

The water bottle holds
1 **liter**. So, the soda bottle
holds more than 1 liter.

The water bottle holds 1 liter.
So, the mug holds less than
1 liter.

A **milliliter** is a very small amount.

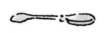

A teaspoon holds about
5 milliliters.

A tablespoon holds about
15 milliliters.

A milliliter is
about 10 drops.

You can measure in liters and milliliters.

EXAMPLE 1 How many milliliters would a drinking glass hold?

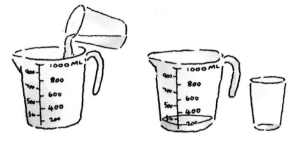

★ **ANSWER** A drinking glass holds about 200 milliliters.

EXAMPLE 2 Could this fish bowl hold 5 liters or 50 liters of water?

Think The bowl holds more than a liter, but *not* 50 times as much.

★ **ANSWER** This fish bowl could hold 5 liters of water.

MATH ALERT

Be Careful When Comparing Capacities

Some containers may look larger than others. But you often have to measure to be sure.

The taller container may look like it holds more than the shorter container. But you can't be sure without measuring.

You use a **thermometer** to measure **temperature** to see how hot or cold something is.

The higher the temperature, the warmer it is.

Fahrenheit

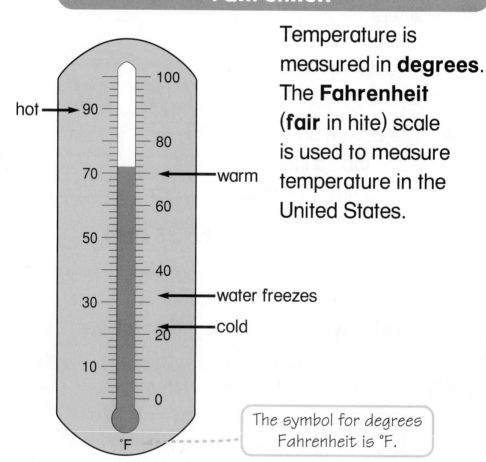

hot → 90

100

80

70 ←warm

60

50

40

30 ←water freezes

20 ←cold

10

0

°F

Temperature is measured in **degrees**. The **Fahrenheit** (**fair** in hite) scale is used to measure temperature in the United States.

The symbol for degrees Fahrenheit is °F.

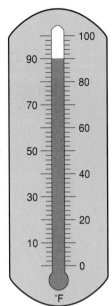

The picture shows a hot day.

The thermometer shows 90°F.

Say   *ninety degrees Fahrenheit*

The picture shows a cold day.

The thermometer shows 24°F.

Say   *twenty-four degrees Fahrenheit*

Celsius

The **Celsius** (C) scale is used to measure
temperature.

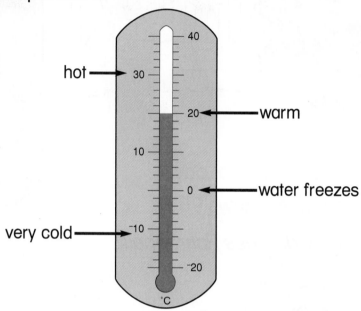

hot — 30

20 — warm

10

0 — water freezes

very cold — ⁻10

⁻20

°C

The picture shows a
comfortable room.

The thermometer
shows 20°C.

Say   *twenty degrees Celsius*

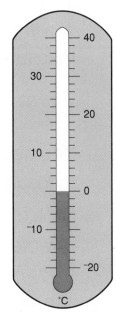

The picture shows a cold day.

The thermometer shows 0°C.

Say *zero degrees Celsius*

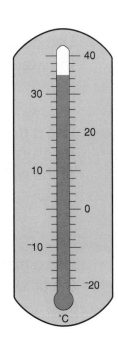

The picture shows a hot day.

The thermometer shows 35°C

Say *thirty-five degrees Celsius*

You can measure one thing in many different ways. Use the right tools and the right units to measure what you need to find out.

Measurement Tools

You need to use the right tool to find the measurements you want.

Look at how you can measure a pitcher of water.

More Help
See
210,
218,
222,
226

Use a ruler to measure the height of the water.

Use a scale to measure the weight of the water and the pitcher.

Use a thermometer to measure the temperature of the water.

Use a measuring cup to measure the amount of water in the pitcher.

Measurement Units

Sometimes you have to choose *which size* unit to use. Sometimes you have to choose *which kind* of unit to use.

EXAMPLE 1 Would you use inches or yards to measure the width of your hand?

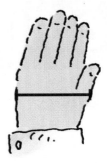

> Measure small things with small units.

More Help See 210–211

★ **ANSWER** inches

EXAMPLE 2 Is the capacity of a pot 5 liters or 5 meters?

> Measure *capacity* with liters.
> Measure *length* with meters.

More Help See 213, 222, 224

★ **ANSWER** 5 liters

More Help See 211, 218

EXAMPLE 3 Could the length of a path be 10 yards or 10 pounds?

> Measure *length* in yards.
> Measure *weight* in pounds.

★ **ANSWER** 10 yards

Graphing, Statistics, and Probability

You can learn a lot by collecting information. You can learn what food people like best. You can learn to predict the time when a bus will go by your school.

Using a table or chart can help you keep track of your information. Some kinds of information, however, are easier to collect than others.

Pieces of information are called **data**. Data can be shown in charts, tables, and graphs. The math used with data is called **statistics**.

Tally Charts

Tally marks are an easy way to keep track of data you collect in a **survey**. You make 1 mark for each answer counted.

/	//	///	////	⊬⊢⊢	⊬⊢⊢ /
1	2	3	4	5	6

> When you get to 5, draw a slash through the first 4 marks.

More Help
See 60, 97

You can use a **tally chart** when you collect data.

You can count the groups of 5 tallies by fives. Then count single tallies by ones.

What is in your lunch bag?

Food	Tally	Number
Sandwich	⊬⊢⊢ ⊬⊢⊢ ⊬⊢⊢ //// 5 10 15 16 17 18 19	19
Apple	⊬⊢⊢ /// 5 6 7 8	8
Banana	/// 1 2 3	3
Orange	⊬⊢⊢ / 5 6	6

Tables

A **table** shows data in rows and columns.
It makes it easy to find and
compare data.

Do you like smooth or chunky peanut butter?
Asha — Smooth
Vince — chunky
Danielle — chunky

I like chunky peanut butter.

Here are the two kinds of peanut butter.

Favorite Peanut Butter

	Chunky	Smooth
Boys	2	4
Girls	3	5
Total	5	9

2 boys chose chunky.
4 boys chose smooth.

3 girls chose chunky.
5 girls chose smooth.

5 children chose chunky.
9 children chose smooth.

More of the children like smooth peanut butter than like chunky peanut butter.

Did You Know?

Americans eat a lot of peanuts.
Half of those peanuts are in the
form of peanut butter.

Source: peanut-institute.org/PeanutFAQs

Chunky Chunky

Smooth

Graphs

A **graph** can help you see a picture of information. A graph can make it easy to compare numbers.

Real Graphs

You can make a **real graph** with objects or with models.

EXAMPLE 1 Does Maya have more toy bears or toy rabbits?

More Help
See 26

One Way Use the real objects.

Maya's Toys

Rabbits	🐰	🐰	🐰		
Bears	🐻	🐻	🐻	🐻	🐻

Another Way Use models.

Maya's Toys

Rabbits	▭▭▭
Bears	▭▭▭▭▭

You can use 🧱 for each rabbit. Use 🧱 for each bear.

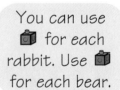

⭐ **ANSWER** Either way, Maya has more toy bears than toy rabbits.

Picture Graphs

A **picture graph** uses pictures to show data.

Favorite Color

Red									
Blue									
Yellow									

Each face stands for a child
who voted for a favorite color.

You can use the graph to answer questions.

EXAMPLE How many more children voted for red
than for blue?

 Subtract to compare.

    red   blue

$$9 - 5 = 4$$

**More
Help**
See
80–
81

 Count to compare.

Red

Blue   4 more

★ **ANSWER** Four more children voted for red
than for blue.

Bar Graphs

> It helps if you know the parts of a graph.

A **bar graph** lets you compare numbers by comparing the lengths of different bars.

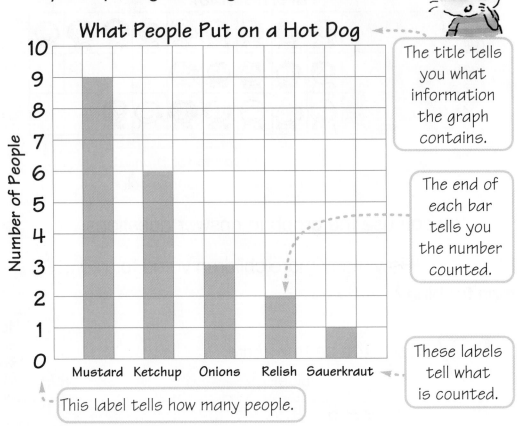

What People Put on a Hot Dog

Number of People

Mustard Ketchup Onions Relish Sauerkraut

The title tells you what information the graph contains.

The end of each bar tells you the number counted.

These labels tell what is counted.

This label tells how many people.

Here are some things this bar graph shows.

- More people use ketchup than use relish.
- Most people use mustard.
- Not very many people use sauerkraut.

There are things the graph does *not* tell you.
It doesn't tell you how many people use ketchup and relish *together*.

You can make a bar graph.

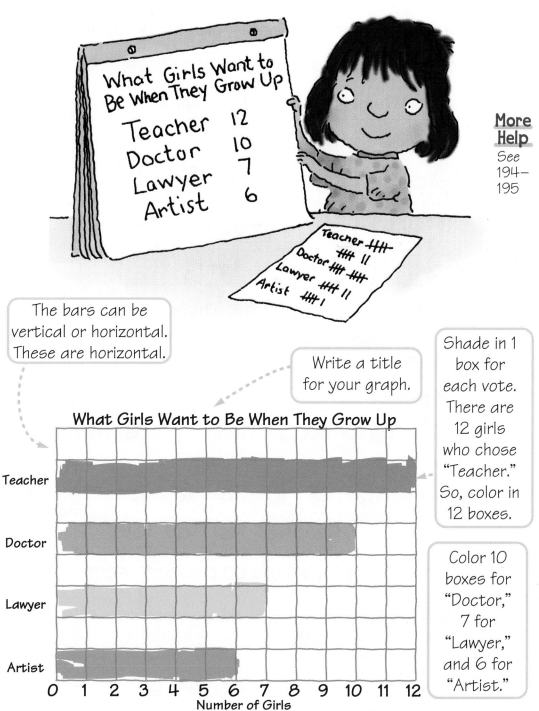

What Girls Want to Be When They Grow Up

Teacher 12
Doctor 10
Lawyer 7
Artist 6

Teacher ‖‖‖‖ ‖‖‖ ‖‖
Doctor ‖‖‖‖ ‖‖‖‖‖
Lawyer ‖‖‖‖ ‖‖
Artist ‖‖‖‖ ‖

More Help
See 194–195

The bars can be vertical or horizontal. These are horizontal.

Write a title for your graph.

Shade in 1 box for each vote. There are 12 girls who chose "Teacher." So, color in 12 boxes.

What Girls Want to Be When They Grow Up

Teacher	
Doctor	
Lawyer	
Artist	

0 1 2 3 4 5 6 7 8 9 10 11 12
Number of Girls

Color 10 boxes for "Doctor," 7 for "Lawyer," and 6 for "Artist."

Pictographs

In a **pictograph,** symbols stand for numbers.

EXAMPLE How many players are on a soccer team?

Number of Players on Teams

Ice Hockey	🙂 🙂 🙂
Softball	🙂 🙂 🙂 🙂 🙂
Soccer	🙂 🙂 🙂 🙂 🙂 🙂

Key 🙂 = 2 players

Every pictograph has a **key**. The key tells you what each symbol stands for. On this graph, each 🙂 stands for 2 players. So, a 🙂 stands for 1 player.

Find the row for *Soccer*. Count by twos. Then add 1 more.

Soccer

There are $5\frac{1}{2}$ faces.

⭐ **ANSWER** 11 players are on a soccer team.

Did You Know?

Soccer is the most popular sport in the world. It is played in more than 140 countries.

Source: *The Information Please® Kids' Almanac,* Houghton Mifflin Company

You can make a pictograph.

Cutest Baby Animal

🐯	🧍🧍
🐕	🧍🧍🧍🧍🧍
🦘	🧍🧍🧍
🦭	🧍🧍🧍

Key 🧍 = 10 votes

Write a title for your graph.

Choose a symbol to stand for the votes. Decide on a value for each symbol. The votes seem to be in groups of 5. So, 5 or 10 would be a good choice.

Write a key that shows what each symbol stands for.

Did You Know?

Animal groups have special names.

a pride of lions   a troop of kangaroos
a herd of elephants   a pod of seals

Source: *The Teacher's Book of Lists*, Scott, Foresman and Company

Venn Diagrams

More Help
See
196

You can sort things into groups. A **Venn diagram** uses shapes like circles or rings to show the groups.

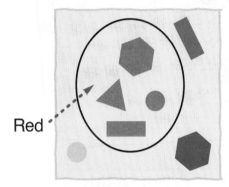

Red

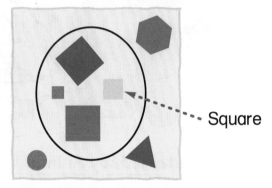

Square

The ring shows which blocks are in the "red" group.

The ring shows which blocks are in the "square" group.

The blocks outside the ring are not red.

The blocks outside the ring are not square.

More Help
See
196–197

A Venn diagram may have more than one ring.

The hexagons and circles do not fit in either group.

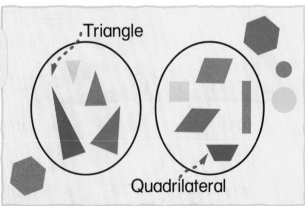

Triangle

Quadrilateral

Sometimes one thing fits in more than one group. That thing goes inside an area where the loops overlap.

The red triangles belong in both groups.

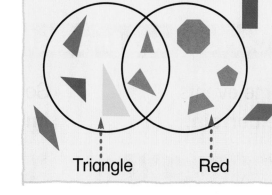

Triangle   Red

More Help
See 196

You can sort just about anything.

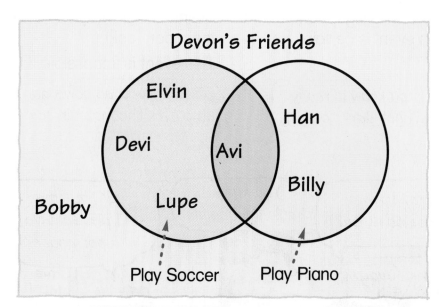

Devon's Friends

Elvin

Han

Devi   Avi

Billy

Bobby   Lupe

Play Soccer   Play Piano

Bobby doesn't play soccer or the piano. Avi plays both.

When you talk about whether something might happen, you are talking about its **probability**.

An **event** is something that *may* happen or *may not* happen.

Certain and Impossible Events

- Some events are **certain**.
- Some events are **impossible**.

Event: It will get dark tonight. This event is certain.

Event: A cow will jump over the moon tonight.
This event is impossible.

> You *know* it really will get dark tonight.

> You *know* that cows don't jump over the moon in real life.

EXAMPLE 1 You pick a cube from this bag. Is it *certain* or *impossible* that the cube is red?

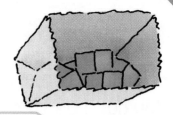

> All the cubes are red. It would be impossible to pick a cube that is *not* red.

The only color you can pick is red.

★ **ANSWER** It is certain that the cube is red.

EXAMPLE 2 You pick a cube from this box. Is it *certain* or *impossible* that the cube is red?

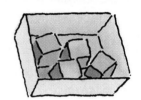

> None of the cubes are red. It would be impossible to pick a red cube.

★ **ANSWER** It is impossible to pick a red cube.

EXAMPLE 3 You roll a number cube that has faces marked 1, 2, 3, 4, 5, and 6. Is it *certain* or *impossible* that you roll an 8?

More Help
See 205

> There is no 8 on the cube.

★ **ANSWER** It is impossible to roll an 8.

> If you had a cube with 8 on each face, then it would be certain that you would roll an 8.

Likely and Unlikely Events

Sometimes an event *might* happen.
You can't be sure.

- An event can be **likely**.

It probably will rain, but it's not certain. It could be cloudy all day and not rain.

Event: It will rain today.
The picture shows that this event is likely.

More Help
See
244

- An event can be **unlikely**.

It probably won't rain, but it's *not* impossible.

Event: It will rain today.
The picture shows that this event is unlikely.

EXAMPLE 1 You spin this spinner. Is it *likely* or *unlikely* that the arrow will land on red?

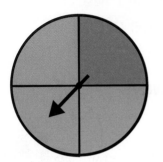

The spinner has 4 equal parts. 3 are red and 1 is blue. There is more red than blue on the spinner.

More Help See 42

The arrow could land on blue, but it's more likely that it will land on red.

★ **ANSWER** It is likely the arrow will land on red.

EXAMPLE 2 You spin this spinner. Is it *likely* or *unlikely* that the arrow will land on yellow?

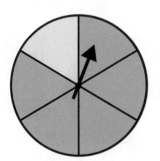

The spinner has 6 equal parts. 1 is yellow and 5 are purple. There is more purple than yellow on the spinner.

The arrow could land on yellow, but it is more likely it will land on purple.

★ **ANSWER** It is unlikely the arrow will land on yellow.

Equally Likely Events

Some events are **equally likely.** That means that they each have the same chance of happening.

If you flip a penny, there are two ways it could land.

heads   tails

Every time you flip a coin there are two events that can happen. The coin can land heads up. The coin can land tails up. The two events are equally likely.

You could flip a coin 50 times and count the number of heads and tails. The numbers probably won't be the same, but they usually will be close.

EXAMPLE 1 You spin this spinner.
The arrow may land on red.
The arrow may land on blue.
Are these events equally likely?

More Help See 42
The chances of the arrow landing on red and the chances of the arrow landing on blue are the same.

The spinner has two equal parts. 1 is red and 1 is blue.

★ **ANSWER** Yes, the events are equally likely.

EXAMPLE 2 You and a friend want to play this spinner game.

> ### Spinner Game
> #### Rules
>
> - Choose a spinner with two colors.
> - You and a friend each choose a different color from the spinner.
> - Take turns spinning the spinner.
>   If the arrow lands on your color, you get a point.
>   If the arrow lands on your friend's color, your friend gets a point.
> - Keep spinning until you each have 10 turns.
> - The player with more points wins.

You want the game to be fair. Which spinner would you choose?

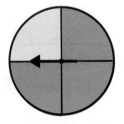

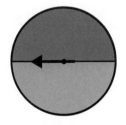

Red is more likely than yellow.   Blue and orange are equally likely.

★ **ANSWER** Use the blue and orange spinner for a fair game.

Making Predictions

You make a **prediction** when you use what you already know to tell what you think will probably happen next.

Sometimes you make predictions during an experiment.

1. Predict what you will pull out of the bag.

   That's a silly guess.

   A duck?

2. Take 1 thing out of the bag. Then put it back.

3. Repeat 100 times. Record your results.

   Red ball: 90 times
   Blue ball: 10 times

4. Now predict what you will pull out of the bag.

   Red ball

"Red ball" is a good prediction.

EXAMPLE There are 85 children in grades 1 and 2. Use the information below to predict if more children will bring their lunch or buy a school lunch tomorrow.

Will you bring lunch or buy lunch tomorrow?
Bring lunch ~~IIII~~ ~~IIII~~ II
Buy lunch III

12 children said they would bring lunch.

3 children said they would buy lunch.

You didn't ask all the children in grades 1 and 2. But you can use the information from the children you did ask to make a prediction.

More Help See 234

★ **ANSWER** Saying that more children would bring lunch tomorrow is a good prediction.

It is possible that more children will actually buy lunch tomorrow, but it isn't likely. A good prediction is just the best guess you can make with what you know.

Algebraic Thinking

We should
have 10
animals,
but we only
have 8.

Suppose you are 7 years old and your friend is 8 years old. When you are 8, your friend will be 9. When you are 9, your friend will be 10. Your ages will change every year, but your friend will always be one year older than you are.

Thinking about patterns and about how numbers work is called algebraic thinking.

When something happens over and over, we say there is a **pattern**. You can find patterns in music, art, science, and in math.

More Help
See 276–277

Patterns

There are many different kinds of patterns.

- A pattern can be a group of items that repeats over and over in a row.

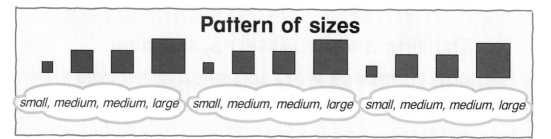

Pattern of sizes

small, medium, medium, large small, medium, medium, large small, medium, medium, large

Pattern of letters

A, B, B, C, A, B, B, C, A, B, B, C

Pattern of sounds

clap, snap, snap, slap   clap, snap, snap, slap   clap, snap, snap, slap

These are three different ways to show the same pattern.

- A pattern can be a group of items that repeats over and over on a grid.

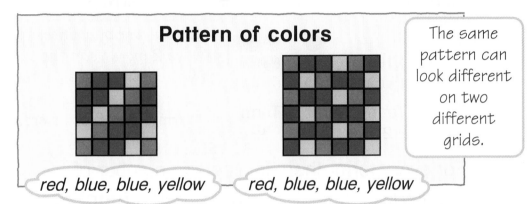

Pattern of colors

The same pattern can look different on two different grids.

red, blue, blue, yellow — *red, blue, blue, yellow*

- A pattern can grow.

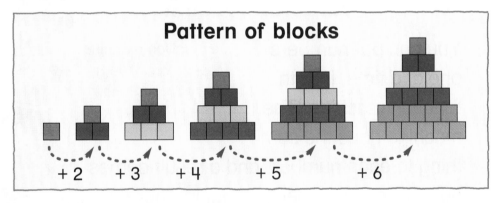

Pattern of blocks

$+2$ $+3$ $+4$ $+5$ $+6$

Pattern of numbers

1   3   6   10   15   21

$+2$ $+3$ $+4$ $+5$ $+6$

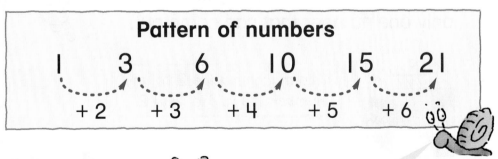

The blocks and the numbers show the same pattern.

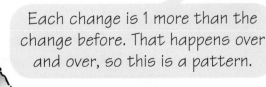

Each change is 1 more than the change before. That happens over and over, so this is a pattern.

Functions

Here are the first four terms in a pattern.

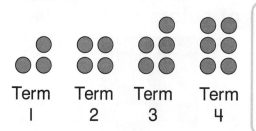

Term 1 Term 2 Term 3 Term 4

> The number of counters is always 2 more than the term. If you continue this pattern, term 5 has 7 counters, term 6 has 8 counters, and so on.

For every term, there is a certain number of counters. This relationship is called a **function.**

Function Machines

You can put numbers, one at a time, into the function machine. The machine does the same thing to each number, and a result comes out. For any number you put in, there's only one number that can come out.

> A function machine isn't real. It is a way to show how functions work.

> This machine adds 2 to any number. Its rule is "add 2."

> This machine subtracts 1 from any number. Its rule is "subtract 1."

Function Tables

A function table shows what goes in and what comes out of a function machine. It shows how the rule relates the "In" number to the "Out" number.

More Help
See 101, 235 282–284

Rule: − 2	
In	Out
2	0
3	1
4	2
5	3

In	Rule	Out
2 − 2	=	0
3 − 2	=	1
4 − 2	=	2
5 − 2	=	3

This function table also shows a pattern in each column. That's because the "In" numbers are in counting order.

EXAMPLE Complete the function table.

Rule: + 2	
In	Out
7	■
■	5
6	■
■	7

In	Rule	Out
7 + 2	=	9
3 + 2	=	5
6 + 2	=	8
5 + 2	=	7

Rule: + 2	
In	Out
7	9
3	5
6	8
5	7

More Help
See 32, 84–85

This function table does not show a pattern in the columns. The "In" numbers are not in counting order.

An **equation** is a number sentence that shows that two amounts are **equal**. You use an equal sign in an equation.

The amount on each side of the equal sign must be the same. That makes sense!

$$3 + 2 = 5$$
$$5 = 5$$

Using a Balance to Complete an Equation

An equation must be balanced. There must be the same amount on each side of the equal sign.

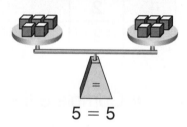

$$5 = 5$$

EXAMPLE 1 $5 + \blacksquare = 5 + 3$

You can use models on a balance.

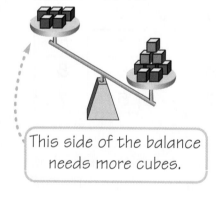

This side of the balance needs more cubes.

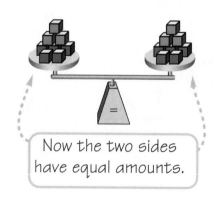

Now the two sides have equal amounts.

$$5 + 3 = 5 + 3$$
$$8 = 8$$

★ **ANSWER** $5 + 3 = 5 + 3$

EXAMPLE 2 $10 = 4 + \blacksquare$

This side of the balance
needs more cubes.

Now the two sides
have equal amounts.

$$10 = 4 + 6$$
$$10 = 10$$

★ **ANSWER** $10 = 4 + 6$

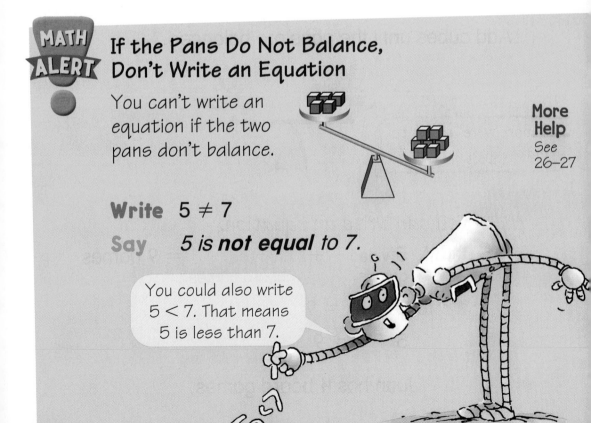

MATH ALERT **If the Pans Do Not Balance, Don't Write an Equation**

You can't write an
equation if the two
pans don't balance.

More
Help
See
26–27

Write $5 \neq 7$

Say *5 is **not equal** to 7.*

You could also write
$5 < 7$. That means
5 is less than 7.

Missing Addend Equations

EXAMPLE 1 Juan has 9 games. Five of the games are video games. The rest are board games. How many board games does Juan have?

More Help See 84–85

One Way You can use models on a balance.

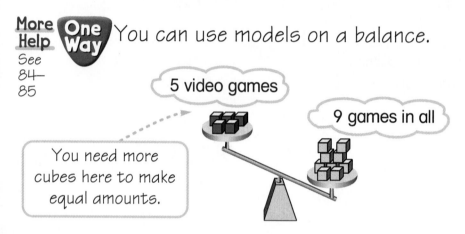

5 video games

9 games in all

You need more cubes here to make equal amounts.

Add cubes until the pans are balanced.

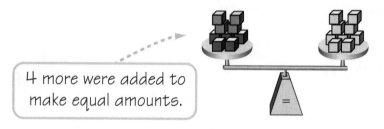

4 more were added to make equal amounts.

Another Way You can write an equation.

Think 5 video games + some board games = 9 games

Write $5 + \blacksquare = 9$

$5 + 4 = 9$

★ **ANSWER** Juan has 4 board games.

EXAMPLE 2 The toy car weighs 4 pounds. How much does the doll weigh?

More
Help
See
218–
219

7 POUNDS

One Way You can use models.

4 for the car.

7 for the total.

More
Help
See
84–
85

Add cubes until you have equal amounts.

4 for the car.
3 for the doll.

7 for the total.

> 4 + 3 = 7
> 7 = 7
> The amounts are equal.

Another Way You can write an equation.

Think 4 pounds + some pounds = 7 pounds
for the car   for the doll

Write 4 + ■ = 7

4 + 3 = 7

★ **ANSWER** The doll weighs 3 pounds.

A **property** is a rule that is true for a whole set of numbers.

Properties of Addition

• Order Property of Addition

More
Help
See
55, 58

You can change the order of addends. The sum does not change.

$$3 + 6 = 9 \qquad 6 + 3 = 9$$

• Grouping Property of Addition

More
Help
See
70–71

You can change the way addends are grouped. The sum does not change.

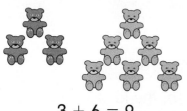

$$4 + 2 + 6 = \blacksquare \qquad 4 + 2 + 6 = \blacksquare$$
$$6 + 6 = 12 \qquad 10 + 2 = 12$$

You can make a 10.

• Adding Zero Property of Addition

More
Help
See
59

When you add zero to a number, the sum is that same number.

$$9 + 0 = 9 \qquad 99 + 0 = 99 \qquad 999 + 0 = 999$$

You'll never change a number by adding zero.

Properties of Multiplication

• Order Property of Multiplication

You can change the order of two factors.
The product does not change.

More Help
See 93, 102–103

$2 \times 3 = 6$ $3 \times 2 = 6$

• Multiplying by 1 Property

When you multiply a number by 1,
the number stays the same.

More Help
See 104

2 groups with 1 bird   1 group with 2 birds
$2 \times 1 = 2$ $1 \times 2 = 2$

> If either factor is 1, the product is the other factor.

• Zero Property of Multiplication

When you multiply a number by 0,
the product is zero.

More Help
See 105

4 groups of 0 birds   0 groups of 4 birds
$4 \times 0 = 0$ $0 \times 4 = 0$

> If either factor is 0, the product is 0.

Coordinate Graphs

A **coordinate graph** is like a map. You can use two numbers in order to tell where a point is on a graph.

- You can use an **ordered pair** to find a point on a graph.

EXAMPLE 1 What animal is at (3, 5)?

Think (3, 5)

The second number tells you how many spaces to move up from 0.

The first number tells you how many spaces to move from 0 to the right.

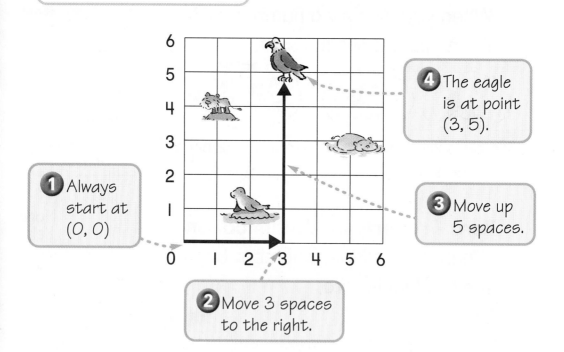

1 Always start at (0, 0)

2 Move 3 spaces to the right.

3 Move up 5 spaces.

4 The eagle is at point (3, 5).

★ **ANSWER** The eagle is at (3, 5).

- You can use an ordered pair to name a point on a graph.

EXAMPLE 2 What ordered pair tells where the duck is?

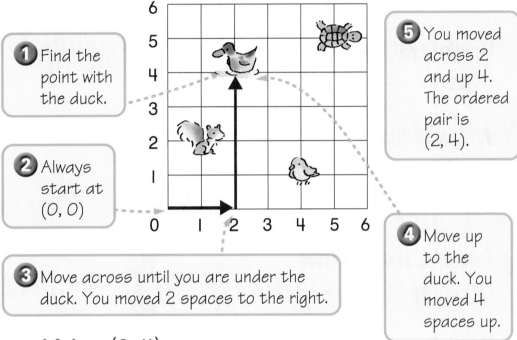

1 Find the point with the duck.

2 Always start at (0, 0)

3 Move across until you are under the duck. You moved 2 spaces to the right.

4 Move up to the duck. You moved 4 spaces up.

5 You moved across 2 and up 4. The ordered pair is (2, 4).

Write (2, 4)

Say *point two four*
or *ordered pair two four*

★ **ANSWER** The duck is at (2, 4).

Did You Know?

Road maps use coordinate grids to make it easier to find places. They use numbers and letters, not just numbers.

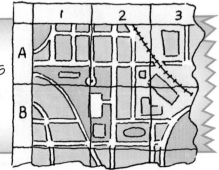

Problem Solving

You solve problems every day. Sometimes you remember what to do. Other times you have to think of new ideas. That's also how you solve math problems. Just be sure to use good ideas and don't blame the dog.

Keep trying.

You can't solve every problem the first time.
That's OK. Look at what you did. Then try again.

Practice.

There is a saying that practice makes perfect.
Practice may not make you perfect, but it does
help you do better. That's true in math, too.
When you try to solve problems, you learn new
things. You become better at solving problems.

Use what you know.

Think back. Remember a lesson you've learned. This works with problem solving, too. Think about other problems you have solved.

Take a break.

Some problems can be hard. You may need to do something else for a while. Then go back to the problem.

Did you ever get stuck on a problem? It can feel like you are lost. Here is a plan that is like a map. It can help you find your way.

❶ Understand

Read the problem carefully. Ask yourself:
- What do I know?
- What do I need to find out?

❷ Plan

Think about what you can do. There are strategies you can use. Can you act out the problem? Can you draw a picture or make a table? Pick a strategy and try it.

Look on page 271 to find more strategies you can use.

❸ Try

Go ahead with the strategy you picked.
If it doesn't work, try a different strategy.

❹ Look Back

When you find an answer, don't stop. Ask:
- Does it answer the question?
- Did I compute correctly?
- Does my answer make sense?

Solving a problem is like going somewhere. There are many ways to travel. There are also many ways to get to an answer. These ways are called problem-solving strategies.

Act It Out

Draw a Picture

Find a Pattern

Guess, Check, and Revise

Make a List

Make a Table

Here are some things to remember when choosing strategies.

- You can use more than one strategy on a single problem.
- You can make up your own strategies.
- You can use one strategy and your friend can use a different one.

Act It Out

You can use objects to act out a problem. You can use counters or coins. You can use cubes, marbles, or paper. You can use anything that is handy.

EXAMPLE The first row has 7 seats. The second row has one more seat than the first row. How many seats are in the first two rows?

➊ Understand

- **What do you know?**
  One row has 7 seats.
  Another row has 1 more than that.

- **What do you need to find out?**
  the number of seats in two rows

➋ Plan

You can act out the problem.
Use cubes to stand for seats.

❸ Try

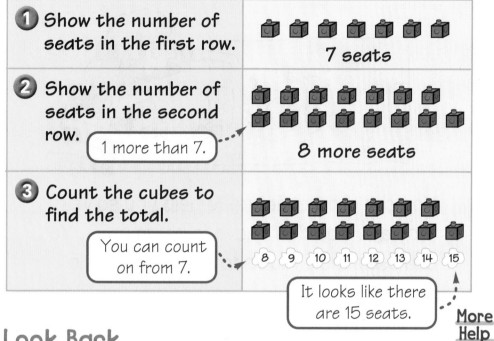

❶ Show the number of seats in the first row.	7 seats
❷ Show the number of seats in the second row. *1 more than 7.*	8 more seats
❸ Count the cubes to find the total. *You can count on from 7.*	8 9 10 11 12 13 14 15

It looks like there are 15 seats.

More Help

See 60

❹ Look Back

- Are there 2 rows? Yes.
- Is 8 one more than 7? Yes.
- Does 7 + 8 = 15? Yes.
- Does my answer make sense? Yes.

★ **ANSWER** There are 15 seats in the first two rows.

Draw a Picture

You can draw a picture to help you solve a problem. You don't have to be an artist. You can use simple marks to stand for people and things.

EXAMPLE You are forming teams. Each team needs 6 players. There are 22 children who want to play. How many teams can you make?

① Understand

● **What do you know?**

More Help See 110

Each team needs 6 players.
You have 22 children.

● **What do you need to find out?**

how many teams you can make

② Plan

You can draw a picture.

❸ Try

Draw an X for each player. Find how many groups of 6 you can make.

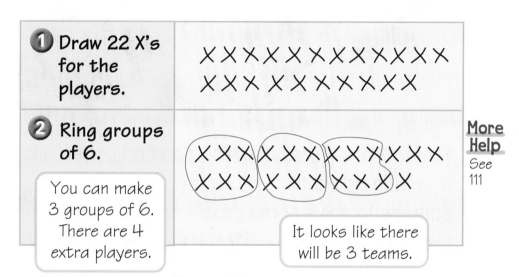

❶ Draw 22 X's for the players.	X X X X X X X X X X X X X X X X X X X X
❷ Ring groups of 6. You can make 3 groups of 6. There are 4 extra players.	X X X X X X X X X X X X X X X X X X X X X X It looks like there will be 3 teams.

More Help
See 111

❹ Look Back

- Did you draw one X for each player? Yes.
- Does each ring have 6 X's in it? Yes.

★ **ANSWER** You can form 3 teams with 6 players each.

There are 4 players left over. Not enough for another team.

The picture doesn't show any children. That's OK. This kind of picture is just to help you solve the problem.

Find a Pattern

More Help
See 254–255

Patterns are all around. You can find patterns in music, in art, and in nature. You can use patterns to solve some problems.

EXAMPLE Find the next two items in each pattern.

Pattern 1: ■■●■■●■ __ __

Pattern 2: ←↑→↓←↑→↓←↑→ __ __

① Understand

- What do you know?
  There are 2 different patterns.

- What do you need to find out?
  the next two items in each pattern

② Plan

Look for things that happen over and over.
Use each pattern to decide what comes next.

❸ Try

Pattern 1: There are 2 squares and then
1 circle. That happens twice and starts again.
The pattern is *square, square, circle.*

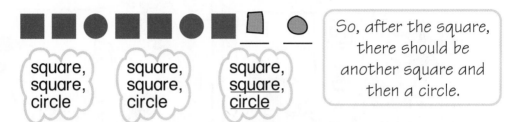

square,
square,
circle

square,
square,
circle

square,
<u>square,</u>
<u>circle</u>

So, after the square,
there should be
another square and
then a circle.

Pattern 2: The arrows point left, up, right, down.
This repeats. The pattern is *left, up, right, down.*

More
Help
See
300

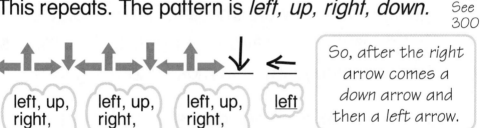

left, up,
right,
down

left, up,
right,
down

left, up,
right,
<u>down</u>

<u>left</u>

So, after the right
arrow comes a
down arrow and
then a left arrow.

❹ Look Back

• Check to be sure your answers fit the pattern.

★ ANSWER

Pattern 1: ■ ■ ● ■ ■ ● ■ ▢ ◯

Pattern 2: ← ↑ → ↓ ← ↑ → ↓ ← ↑ → ↓ ←

Guess, Check, and Revise

I wish I could remember your shoe size. I'll just guess and then check to see if the shoes fit.

These are too big. Let's try a smaller size.

To solve some problems, you can start by guessing an answer. Check it. If it works, keep it. If it doesn't work, revise your guess. That means use what you learned to make another guess.

EXAMPLE There are 8 riding toys. They have 19 wheels in all. How many are scooters and how many are tricycles?

1 Understand

- **What do you know?**
  Scooters have 2 wheels.
  Tricycles have 3 wheels.
  There are 8 riding toys. There are 19 wheels.

- **What do you need to find out?**
  how many scooters there are
  how many tricycles there are

❷ Plan

Use Guess, Check, and Revise for a problem like this.

More Help See 96

❸ Try

Guess.	Check.	Revise.
4 scooters 4 tricycles 8 toys	2 + 2 + 2 + 2 = 8 wheels 3 + 3 + 3 + 3 = 12 wheels 8 + 12 = 20 wheels	Too many wheels! Try fewer tricycles.
6 scooters 2 tricycles 8 toys	2 + 2 + 2 + 2 + 2 + 2 = 12 wheels 3 + 3 = 6 wheels 12 + 6 = 18 wheels	Not enough wheels! Try more tricycles.
5 scooters 3 tricycles 8 toys	2 + 2 + 2 + 2 + 2 = 10 wheels 3 + 3 + 3 = 9 wheels 10 + 9 = 19 wheels	Just right!

It looks like there are 5 scooters and 3 tricycles.

❹ Look Back

- Are there 19 wheels in all? Yes, 10 wheels on scooters and 9 wheels on tricycles.

- Are there 8 riding toys in all? Yes, 5 + 3 = 8.

★ **ANSWER** There are 5 scooters and 3 tricycles.

Make a List

A list can help you keep track of things. That's why making a list can help you solve some kinds of problems.

EXAMPLE Look at the Color-Bots kit. How many different ways can you make a robot?

Heads

Feet

1 Understand

- **What do you know?**

  The head of the robot can be red, blue, or yellow.

  The feet of the robot can be red, blue, or yellow.

- **What do you need to find out?**

  how many different ways you can make a robot

❷ Plan

Try making a list using a pattern. That way you won't miss any robots or list the same robot twice.

More Help
See 276–277

❸ Try

List one robot to start.   red head – red feet

List others with the same color head, red.
red head – blue feet
red head – yellow feet

You could write R, B, and Y instead of red, blue, and yellow.

List robots with a blue head.
blue head – red feet
blue head – blue feet
blue head – yellow feet

List robots with a yellow head.
yellow head – red feet
yellow head – blue feet
yellow head – yellow feet

It looks like there are 9 different ways to make a robot.

You could also make a picture list.

❹ Look Back

- Were any robots left out? No.
- Were any robots listed twice? No.

The pattern makes checking easy.

★ **ANSWER** You can make a robot 9 different ways.

Make a Table

Sometimes you can make a table to help you solve a problem.

EXAMPLE 1 Rona is 7 years old. Her brother is 4. How old will he be when she is 11 years old?

1 Understand

- **What do you know?**

  Rona is 7.

  Her brother is 4.

- **What do you need to find out?**

  Rona's brother's age when she is 11

2 Plan

Make a table to show each child's age each year.

Next year I'll be 1 year older.

I'll be 1 year older, too.

I'll always be 3 years older than you are.

❸ Try

Make a table. Write in the information you know.

Kim's Age	7			
Brother's Age	4			

More Help
See 257

Fill in the table. Keep adding one year to each child's age.

> Stop when Kim's age is 11.

Kim's Age	7	8	9	10	11
Brother's Age	4	5	6	7	8

> It looks like Kim's brother will be 8 years old when Kim is 11.

❹ Look Back

- Does your answer make sense? Yes.

  Kim's age is always 3 more than her brother's age. 11 is 3 more than 8.

★ **ANSWER** When Kim is 11, her brother will be 8.

Happy Birthday Kim

MORE ▶

MORE ON **Make a Table**

EXAMPLE 2 Yuki is making starfish. She needs
5 straws for each starfish. She has
32 straws. How many starfish can she make?

1 Understand

- **What do you know?**

More
Help
See
100–
101

  Each starfish uses 5 straws. Yuki has
  32 straws.

- **What do you need to find out?**

  how many starfish Yuki can make

2 Plan

Make a table. Use a pattern to fill it in.

3 Try

Starfish	1	2	3	4	5	6	7
Straws	5	10	15	20	25	30	35

It looks like Yuki has enough
straws to make 6 starfish.

She does not have
enough straws to
make 7 starfish.

4 Look Back

- **Did you fill in the table correctly?**
  You could use models to check.

★ **ANSWER** Yuki can make 6 starfish.

- You need different skills to do different things.

- You also need skills to solve problems.

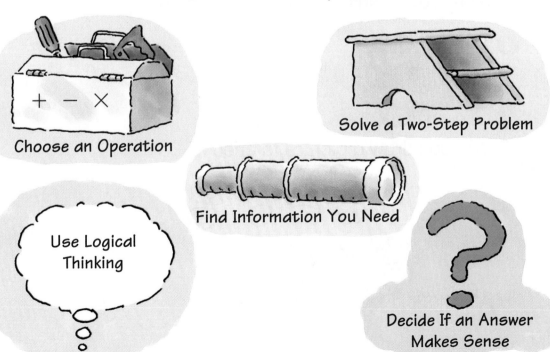

Choose an Operation

Solve a Two-Step Problem

Find Information You Need

Use Logical Thinking

Decide If an Answer Makes Sense

Choose an Operation

Sometimes you can add or subtract to solve a problem.

EXAMPLE 1 Al has 5 cows. Pablo gives him 3 cows. How many cows does Al have now?

① Understand

- What do you know?

More
Help
See
54–
55

  Al has 5 cows. He gets 3 more.

- What do you need to find out?

  how many cows altogether

② Plan

One group joins another. You can add.

③ Try

$5 + 3 = 8$

> 5 cows and 3 more. It looks like Al has 8 cows now.

④ Look Back

More
Help
See
57

You can draw a picture to check.

★ **ANSWER** Now Al has 8 cows.

EXAMPLE 2 There are 7 cows in the barn. 3 cows walk out. How many cows are in the barn now?

① Understand

- What do you know?

  There are 7 cows and 3 leave.

- What do you need to find out?

  how many cows are left

② Plan

Part of a group is taken away.

You can subtract to find how many are left.

More
Help
See
72–73

③ Try

$7 - 3 = 4$ ◄----- 7 cows and 3 leave. It looks like there are 4 cows left in the barn.

④ Look Back

You can act out the problem to check.

More
Help
See
74

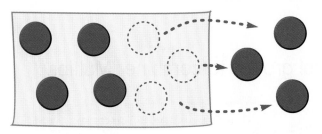

★ **ANSWER** Now there are 4 cows in the barn.

MORE ▶

MORE ON **Choose an Operation**

EXAMPLE 3 You take 3 rides on every roller coaster at Six Flags New England. How many rides is that?

More
Help
See
235

Park	Number of Roller Coasters
Busch Gardens (Virginia)	5
Cedar Point (Ohio)	14
Six Flags New England (Massachusetts)	6

Source: ultimaterollercoaster.com

① Understand

- **What do you know?**

  You ride 6 roller coasters 3 times.

More
Help
See
92

- **What do you need to find out?**

  how many rides you take

② Plan

There are equal groups to combine. Multiply.

③ Try

$6 \times 3 = 18$ ◄ - - - - - It looks like you take 18 roller coaster rides.

④ Look Back

More
Help
See
96

Add to check. $3 + 3 + 3 + 3 + 3 + 3 = 18$

★ **ANSWER** You take 18 roller coaster rides.

EXAMPLE 4 You ride all the roller coasters at Cedar Point. Your cousin rides all the roller coasters at Busch Gardens. How many more coasters do you ride than your cousin does?

① Understand
- What do you know?
  Cedar Point has 14 roller coasters. Busch Gardens has 5 roller coasters.
- What do you need to find out?
  how many more coasters are at Cedar Point than Busch Gardens

② Plan
You can subtract to compare two groups.

More Help
See 80–81

③ Try
$14 - 5 = 9$

It looks like you ride 9 more roller coasters than your cousin does.

④ Look Back
You can check subtraction by adding.
$9 + 5 = 14$

More Help
See 157

★ **ANSWER** You ride 9 more roller coasters than your cousin does.

Solve a Two-Step Problem

Sometimes you need more than one step to solve a problem.

EXAMPLE You have a quarter. You want to buy a monkey and tiger. How much money will you have left?

Lion   15¢
Tiger   10¢
Monkey 5¢
Elephant
         20¢

❶ Understand

- What do you know?

  You have 25¢.

  More Help See 164

  It costs 5¢ for a monkey.

  It costs 10¢ for a tiger.

- What do you need to find out?

  how many cents you will have left if you buy both

❷ Plan

Step 1: Find the total cost for the two items.

Step 2: Find out how much money you will have left.

❸ Try

Step 1: Add to find the total cost.

$$5¢ + 10¢ = 15¢$$

| cost of monkey | cost of tiger | You spend 15¢ in all. |

Step 2: Subtract to find how much is left.

$$25¢ - 15¢ = 10¢$$

| money you began with | money you spent | You will have 10¢ left. |

❹ Look Back

Try a different set of steps to check.

Step 1: Subtract the cost of the monkey from 25¢.

$$25¢ - 5¢ = 20¢$$

You have 20¢ left.

Step 2: Subtract the cost of the tiger from 20¢.

$$20¢ - 10¢ = 10¢$$

You have 10¢ left.

★ **ANSWER** You will have 10¢ left.

More Help See 174, 176

Use Logical Reasoning

Do you ever think like a detective to figure things out? You can do that to solve math problems, too.

EXAMPLE Which card is the mystery card?

Clue 1: It has a number less than 3.

Clue 2: The color of the card is not a
color on the United States flag.

Clue 3: The card has an odd number.

❶ Understand

- **What do you know?**

   The mystery card has a number less than 3.

   The mystery card is *not* red, white, or blue.

   The mystery card has an odd number.

- **What do you need to find out?**

   which card fits all 3 clues

② Plan

Use one clue at a time to take away cards.
Make a picture to keep track.

More Help
See
26,
36–37

③ Try

Clue I says the mystery card has a number less
than 3. So, cross out all the cards with 3.

Clue 2 says the mystery card is not red, white,
or blue. Cross out the red and blue cards.

Clue 3 says the mystery card has an odd
number. Cross out the card with an even number.

The only card left is 1.

④ Look Back

- Is I less than 3? Yes.
- Is the card a color not on the U.S. flag? Yes.
- Is I an odd number? Yes.

★ **ANSWER** The mystery card is 1.

Find Information You Need

Sometimes, there is too much information. You have to find the information you need.

EXAMPLE In 1795, the United States flag had 15 stripes and 15 stars. Today's flag has 13 stripes and 50 stars. Seven of these stripes are red. The rest of the stripes are white. How many white stripes are on the United States flag today?

Source: *The World Almanac for Kids*, World Almanac Books

> In 1795 the 15 stars stood for 15 states. Today the 50 stars stand for 50 states.

➊ Understand

- **What do you know?**
  information about 2 flags

- **What do you need to find out?**
  the number of white stripes on today's flag

❷ Plan

Circle the information you need.

In 1795, the United States flag had 15 stripes and 15 stars. (Today's flag has 13 stripes) and 50 stars. (Seven of these stripes are red. The rest of the stripes are white.)

Subtract to find the missing number.

❸ Try

$$13 - 7 = 6$$

| number of stripes | number of red stripes | It looks like there are 6 white stripes. |

❹ Look Back

- Did you find the information you needed?
  Yes, the number of stripes today and the number of red stripes.

- What information did you ignore?
  The number of stars and stripes in 1795 and the number of stars today.

- Did you subtract correctly? Yes. $6 + 7 = 13$

★ **ANSWER** There are 6 white stripes on today's United States flag.

Decide If an Answer Makes Sense

Cartoons do not have to make sense.

Answers to problems should make sense.

EXAMPLE Does a total of 58 make sense?

More Help
See 190, 235

Video Store – Cartoons Sold Last Week					
Mon.	Tues.	Wed.	Thurs.	Fri.	Total
8	7	9	8	9	58

❶ Understand

More Help
40–41, 158–159

• What do you need to find out?

if a total of 58 makes sense

❷ Plan

Use rounding to estimate the total.

❸ Try

$8 + 7 + 9 + 8 + 9$

$10 + 10 + 10 + 10 + 10 = 50$

> The total number of cartoons sold is about 50. Yes, 50 is close to 58. But be careful. 58 is closer to 60 than to 50.

❹ Look Back

• Does 58 makes sense? Each of the addends is

More Help
See 28–29

less than 10, so the sum must be less than 50.

So, 58 does not make sense.

⭐ **ANSWER** 58 does not make sense.

Almanac

Numbers

You can use symbols and words for numbers. The symbols are called **numerals**.

Writing Numerals

Look at the dots. They show where to start writing a **numeral**. Then follow the direction of the arrows.

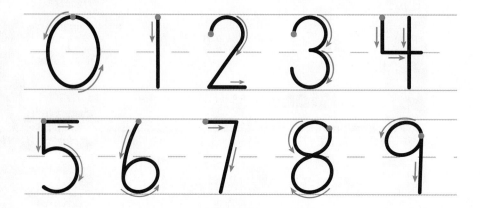

When you write a row of two-digit numbers, you need to leave a space between each number.

The space should be about the width of your finger.

2 4 2 5 2 6 2 7

Word Names for Numbers

0	zero	10	ten	20	twenty
1	one	11	eleven	21	twenty-one
2	two	12	twelve	22	twenty-two
3	three	13	thirteen	23	twenty-three
4	four	14	fourteen	24	twenty-four
5	five	15	fifteen	25	twenty-five
6	six	16	sixteen	26	twenty-six
7	seven	17	seventeen	27	twenty-seven
8	eight	18	eighteen	28	twenty-eight
9	nine	19	nineteen	29	twenty-nine

More Help
See 2, 6–7

Follow this pattern to write the numbers from thirty to ninety-nine.

More Help
See 10–14, 18–23

10	ten	100	one hundred
20	twenty	200	two hundred
30	thirty	300	three hundred
40	forty	400	four hundred
50	fifty	500	five hundred
60	sixty	600	six hundred
70	seventy	700	seven hundred
80	eighty	800	eight hundred
90	ninety	900	nine hundred
		1000	one thousand

Position Words

Left and Right

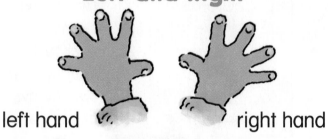

left hand   right hand

The tens digit is on the **left**.

Tens	Ones
2	3

The ones digit is on the **right**.

$$\begin{array}{r} 42 \\ +\ 176 \\ \hline \end{array}$$

These addends are lined up on the right.

Top, Middle, Bottom

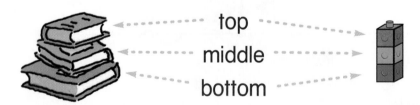

top

middle

bottom

$$\begin{array}{r} 16 \\ 23 \\ +\ 14 \\ \hline \end{array}$$

16 ◄----- top

23 ◄----- middle

14 ◄----- bottom

16 is written above 23.
14 is written below 23.

Before, After, Between

People or things can be in order.

- Alana is standing just **before** Ravi.
- There are two children **after** Suki.
- Sean is **between** Ravi and Suki.

More
Help
See
32-
34

Counting numbers have an order.

- 1 is just **before** 2.
- 7 is just **after** 6.
- 9 is **between** 8 and 10.
- 9 is also **between** 2 and 10.

This page is **between** pages 300 and 302.

Using a Calculator

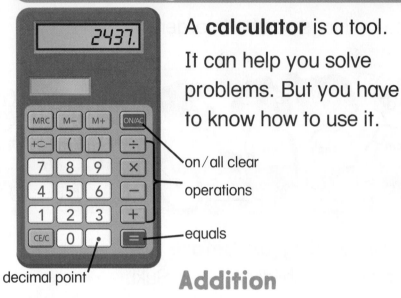

A **calculator** is a tool. It can help you solve problems. But you have to know how to use it.

on/all clear

operations

equals

decimal point

Addition

EXAMPLE 1 47 + 26 = ■

Remember to start by pressing ON/AC .
The display shows 0. .

Press: 4 7 + 2 6 =

See

| 4. | 47. | 47. | 2 | 26. | 73. |

★ **ANSWER** 47 + 26 = 73

EXAMPLE 2 309 + 278 = ■

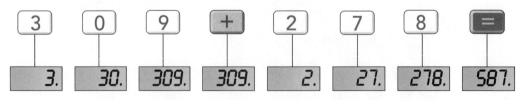

| 3. | 30. | 309. | 309. | 2. | 27. | 278. | 587. |

★ **ANSWER** 309 + 278 = 587

Subtraction

EXAMPLE 1 82 − 38 = ■

Remember to start by pressing ON/AC .
The display shows 0. .

Press: 8   2   −   3   8   =

See   8.   82.   82.   3.   38.   44.

★ **ANSWER** 82 − 38 = 44

EXAMPLE 2 740 − 265 = ■

7   4   0   −   2   6   5   =

7.   74.   740.   740.   2.   26.   265.   475.

★ **ANSWER** 740 − 265 = 475

Did You Know?

For some problems, like 60 + 10 or 138 − 1,
your head is faster than a calculator.

Add and Subtract Money

There is no $ or ¢ on the calculator. But you can use a decimal point to show money amounts.

EXAMPLE 1 $3.95 + $0.97 = ■

Press:

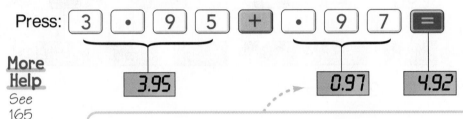

More Help
See
165

> If you don't press a number key before the decimal point, the calculator always puts a 0 there.

★ **ANSWER** $3.95 + $0.97 = $4.92

> Remember to write the $ and . in your answer.

EXAMPLE 2 $9.03 − $3.87 = ■

Press:

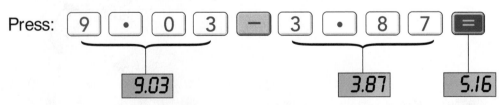

★ **ANSWER** $9.03 − $3.87 = $5.16

> Remember to write the $ and . in your answer.

Skip Counting

You can use a calculator to skip count.

EXAMPLE 1 Skip count by twos from 10.

> Remember to clear the calculator by pressing ⎡ON/AC⎤.

Press:

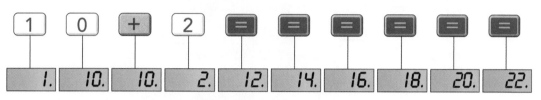

1	0	+	2	=	=	=	=	=	=
1.	10.	10.	2.	12.	14.	16.	18.	20.	22.

★ **ANSWER** 10, 12, 14, 16, 18, 20, 22, ...

EXAMPLE 2 Skip count by fives from 10.

Press:

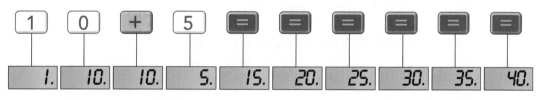

1	0	+	5	=	=	=	=	=	=
1.	10.	10.	5.	15.	20.	25.	30.	35.	40.

★ **ANSWER** 10, 15, 20, 25, 30, 35, 40, ...

EXAMPLE 3 Skip count by tens from 100.

Press:

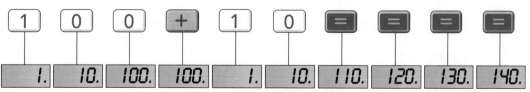

1	0	0	+	1	0	=	=	=	=
1.	10.	100.	100.	1.	10.	110.	120.	130.	140.

★ **ANSWER** 100, 110, 120, 130, 140, ...

Taking a Test

Tests are important. They can help to show what you know. So, you have to learn to be a good test taker.

Test-Taking Tips

- **Wait** until your teacher tells you what to do.
- **Write** your name on your paper.
- **Listen** and follow along as your teacher reads the directions.
- **Ask** questions if you do not understand.
- **Work** each problem in order. Skip any that you are not sure of.
- **Go back** and try to do the problems you skipped.
- **Check** your answers.

Fill-In-the-Blank Test

A fill-in-the-blank test has a space, a line,
or a shape where you write. You may write
numbers, symbols, or words. Sometimes
you will have to draw.

Problems	**Problems with Answers**	

1.
$$\begin{array}{r} 17 \\ + 24 \\ \hline \end{array}$$

1.
$$\begin{array}{r} {\scriptstyle 1} \\ 17 \\ + 24 \\ \hline 41 \end{array}$$

More Help
See 127–129

2. $3 + 4 =$ ___

2. $3 + 4 = \underline{7}$

More Help
See 63

3. Compare.

12 ◯ 15

3. Compare.

12 **<** 15

More Help
See 28–29

4. Draw the next shape.

◯ ▢ ▢ ◯ ◯ ▢ ▢ ◯ __

4. Draw the next shape.

◯ ▢ ▢ ◯ ◯ ▢ ▢ ◯

More Help
See 276–277

5. Complete. Write
odd or *even*.
43 is _____.

5. Complete. Write
odd or *even*.
43 is ___*odd*___ .

More Help
See 36–37

Multiple-Choice Test

More Help
See 60–61
A multiple-choice test lets you choose an answer. Only one answer is correct. So, it helps if you can tell which answers are wrong.

• Pay attention to the operation signs.

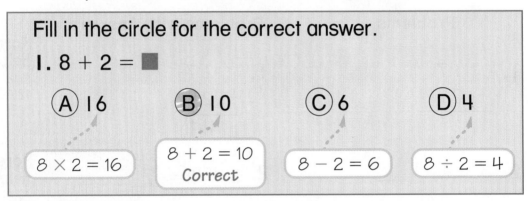

Fill in the circle for the correct answer.

1. $8 + 2 = \blacksquare$

Ⓐ 16   Ⓑ 10   Ⓒ 6   Ⓓ 4

$8 \times 2 = 16$ $8 + 2 = 10$ Correct $8 - 2 = 6$ $8 \div 2 = 4$

• Estimate. Cross out choices that can't be right. Find the exact answer if you need to.

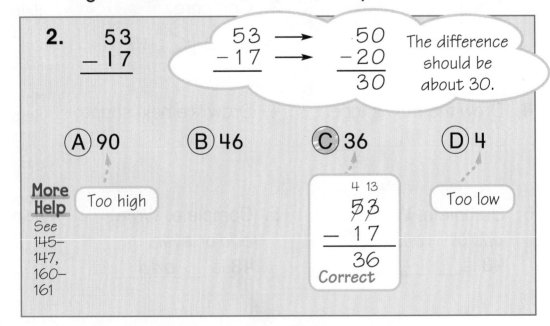

2. 53
       $-\,17$

$53 \longrightarrow 50$
$-\,17 \longrightarrow -20$
$\ 30$

The difference should be about 30.

Ⓐ 90   Ⓑ 46   Ⓒ 36   Ⓓ 4

More Help
See 145–147, 160–161

Too high

Too low

$\overset{4\ 13}{\cancel{5}\cancel{3}}$
$-\ 17$
$\overline{36}$
Correct

Short-Answer Test

Parts of tests may have problems that need a sentence for the answer. These are short-answer tests.

More Help See 80–81, 289

● Write a complete sentence.

Solve the problem. Then check your subtraction. A complete sentence with the wrong answer is still wrong!

1. There are 12 children playing drums and 7 children playing bugles. How many more children play drums than play bugles?

 $\underline{12 - 7 = 5}$

There are 5 more children who play drums than play bugles.

Use words from the question to write your answer.

● Make sure you answer the question.

2. Doug found these coins in his pocket. How much money did he find?

More Help See 169

Doug found 40¢.

The question asks how much money, not how many coins. You need to give the total value of the coins.

Customary Measure

More Help
See 210–211

Length	
1 inch (in.)	——————
1 foot (ft)	12 inches
1 yard (yd)	3 feet
1 mile (mi)	5280 feet

1 inch long

More Help
See 218–219

Weight	
1 ounce (oz)	
1 pound (lb)	16 ounces
1 ton (t)	2000 pounds

More Help
See 222–223

Capacity	
1 teaspoon (tsp)	
1 tablespoon (tbsp)	3 teaspoons
1 fluid ounce (fl oz)	2 tablespoons
1 cup (c)	8 fluid ounces
1 pint (pt)	2 cups
1 quart (qt)	2 pints
1 gallon (gal)	4 quarts

Metric Measure

Length

1 millimeter (mm)	-
1 centimeter (cm)	10 millimeters
1 decimeter (dm)	10 centimeters
1 meter (m)	100 centimeters

1 mm long

1 cm long

More Help See 212–213

Mass

| 1 gram (g) | |
| 1 kilogram (kg) | 1000 grams |

More Help See 220–221

Capacity

| 1 milliliter (mL) | |
| 1 liter (L) | 1000 milliliters |

More Help See 224–225

Roman Numerals

Ancient Romans used letters to write numerals. To write numerals from 1 to 20, they used only three letters.

I = 1   V = 5   X = 10

I	1	XI	11
II	2	XII	12
III	3	XIII	13
IV	4	XIV	14
V	5	XV	15
VI	6	XVI	16
VII	7	XVII	17
VIII	8	XVIII	18
IX	9	XIX	19
X	10	XX	20

Don't use the same letter more than three times in a row.

To read a Roman Numeral, I add the values of the letters.

That doesn't work if the "small" letter comes before the "big" letter. Then you have to subtract.

VIII → 5+1+1+1 = 8
IV → 5−1 = 4

Yellow Pages

A

add Combine two or more groups. **(55)**

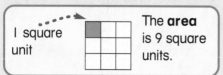

$$4 + 2 = 6$$

addend A number you add. **(55)**

$$5 + 9 = 14$$

addends

Adding Zero Property When you add 0, the sum is the same as the other addend. **(262)**

$$6 + 0 = 6$$

addition The operation of combining groups to find the total amount. **(54)**

after **(33, 301)**

12, 13   12, 13, 14

13 is **after** 12.   14 is also
                after 12.

area The number of square units needed to cover a surface. **(216)**

1 square unit — The **area** is 9 square units.

array A set of objects in equal rows and equal columns. Used to picture multiplication. **(98)**

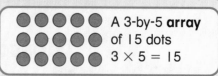

A 3-by-5 **array** of 15 dots
$$3 \times 5 = 15$$

B

balance **(221)**

bar graph **(238)**

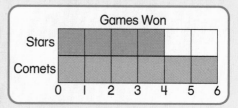

Games Won

before **(33, 301)**

11, 12   10, 11, 12

11 is **before** 12.   10 is also
                before 12.

between **(32, 301)**

18, 19, 20

19 is **between** 18 and 20.

17, 18, 19, 20

18 and 19 are both **between** 17 and 20.

C

calculator (302)

calendar (190)

JANUARY						
S	M	T	W	T	F	S
	1	2	3	4	5	6
7	8	9	10	11	12	13
14	15	16	17	18	19	20
21	22	23	24	25	26	27
28	29	30	31			

capacity The amount a container can hold.
(222, 310, 311)

Celsius (C) The metric scale used to measure temperature.
(228)

cent A unit of money. (164)

100 **cents** = 1 dollar

1 **cent**   5 **cents**

centimeter (cm) A metric unit of length. A **centimeter** is about the width of your finger. (212)

100 **centimeters** = 1 meter

certain Sure to happen. An event that will definitely happen is **certain**. (244)

change Money you get back when you pay for something.
(179)

Have   Buy   **Change**

circle (196)

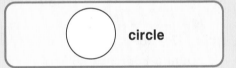

circle

closed figure A figure with all sides connected. (197)

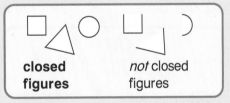

closed figures   *not* closed figures

cone (204)

cone

congruent Same size and same shape. **(200)**

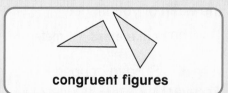

congruent figures

coordinate graph A graph used to show ordered pairs of numbers. **(264)**

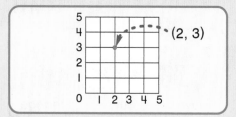

count on A way to add. **(60, 166)**

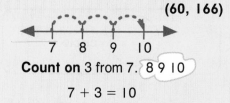

Count on 3 from 7. 8 9 10

7 + 3 = 10

cube A solid figure with six square faces. **(204)**

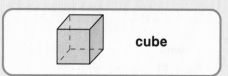

cube

cup (c) A customary unit of capacity equal to 8 ounces. **(222)**

cylinder **(204)**

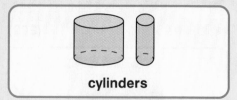

cylinders

D

data Information that has numbers. **(234)**

day A unit of time equal to 24 hours. **(181)**

degrees Units of temperature that are measured in either the Celsius or the Fahrenheit scale. **(226)**

denominator In a fraction, the number of equal parts or groups. The bottom number in a fraction. **(42)**

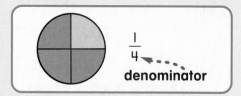

$\frac{1}{4}$ **denominator**

difference The result when one number is subtracted from another. **(73)**

9 − 4 = 5 **difference**

digital clock (180)

digits 0, 1, 2, 3, 4, 5, 6, 7, 8, 9 (8)

dime A coin worth 10 cents. (164)

a **dime**

divide Separate into equal groups. (108)

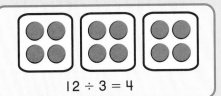

$12 \div 3 = 4$

division The operation of making equal groups to find how many in each group or how many groups. (108)

dollar An amount of money equal to 100 cents. (164)

a **dollar**

double plus one A type of addition fact. (63)

$4 + 4 = 8$ ◀--- double

$4 + 5 = 9$ ◀--- **double plus one**

$4 + 5$ is $4 + 4$ plus 1 more.

doubles Addition facts with two addends that are the same. (62)

$2 + 2 = 4$ $3 + 3 = 6$

elapsed time The amount of time that passes between two times. (188)

equal Is the same amount. (258)

$4 = 3 + 1$

$4 = 4$

4 **equals** $3 + 1$ means that 4 is the same amount as $3 + 1$.

equal parts (42)

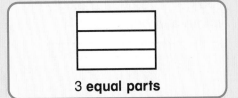

3 **equal parts**

equally likely Having the same chance of happening. **(248)**

For this spinner, spinning red and spinning blue are **equally likely** events.

equation A number sentence with an equal sign. The amount on one side of the equal sign has the same value as the amount on the other side. **(258)**

$$3 + 2 = 5 \qquad 3 + \blacksquare = 5$$

equations

estimate (es´ ti mit) A number close to an exact amount. **(38, 159)**

34 + 19 is about 50.

estimate

estimate (es´ ti mate) To make an estimate. **(38, 158)**

even numbers Numbers that you get when you multiply a whole number by 2. **(36)**

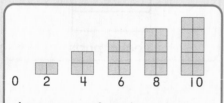

An **even number** always has 0, 2, 4, 6, or 8 in the ones place.

event Something that *will* happen, *may* happen, or *will not* happen. **(244)**

expanded form A way to write numbers that shows the place value for each digit. **(21)**

$$429 = 400 + 20 + 9$$

4 hundreds   2 tens   9 ones

F

face A surface on a solid figure. **(205)**

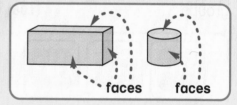

faces   faces

fact family A group of facts that share the same numbers. **(82)**

$$3 + 5 = 8 \qquad 8 - 5 = 3$$
$$5 + 3 = 8 \qquad 8 - 3 = 5$$

factor A number you multiply. **(93)**

$$4 \times 2 = 8$$

factors

Fahrenheit (F) The customary scale used to measure temperature. **(226)**

flip One way to move a figure. **(203)**

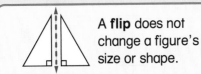

A **flip** does not change a figure's size or shape.

foot (ft) A customary unit of length equal to 12 inches. **(210)**

fourths The parts you get when you divide something into 4 equal parts. **(46)**

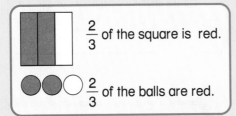

fourths

fraction A way to describe a part of a whole or a group. **(42)**

$\frac{2}{3}$ of the square is red.

$\frac{2}{3}$ of the balls are red.

front-end estimation A way to use front digits to get a close answer. **(159)**

$647 - 266 = \blacksquare$

$600 - 200 = 400$

$647 - 266$ is about 400.

function Pairs of numbers that follow a rule. In a function, there is only one "Out" number for an "In" number. **(256)**

Rule: + 3		In this
In	Out	**function**, the
2	5	rule is "Add 3."
3	6	The "Out"
4	7	number is 3
5	8	more than the
6	9	"In" number.

G

gram (g) A metric unit of mass. **(220)**

1000 **grams** = 1 kilogram

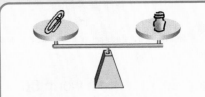

A paper clip has a mass close to 1 **gram**.

graph A picture of data. **(236)**

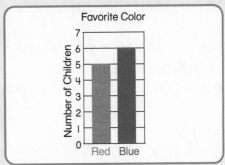

Favorite Color

Number of Children

Red Blue

greater than More than.　(27)

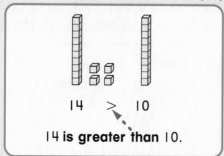

14　>　10

14 **is greater than** 10.

Grouping Property of Addition Changing how you put addends together does not change the sum.　(262)

$$\underline{2 + 3} + 4 = \blacksquare$$
$$5 + 4 = 9$$
$$2 + \underline{3 + 4} = \blacksquare$$
$$2 + 7 = 9$$

 H

half-dollar A coin worth 50 cents.　(165)

a **half-dollar**

halves The parts you get when you divide something into 2 equal parts.　(43)

$\frac{1}{2}$　$\frac{1}{2}$　**halves**

hexagon A figure with 6 straight sides.　(196)

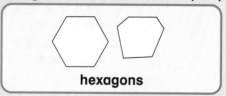

hexagons

horizontal line A line that goes straight across, like the horizon.　(194)

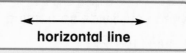

horizontal line

hour (h) A unit of time equal to 60 minutes.　(181)

hour hand The short hand on a clock.　(180)

hour hand

I

impossible An event that will definitely not happen is an **impossible** event.　(244)

inch (in.) A customary unit of length.　(210)

12 **inches** = 1 foot

K

key A sentence that tells what each picture on a pictograph stands for. **(240)**

Key: = 10 votes

kilogram (kg) A metric unit of mass equal to 1000 grams. A **kilogram** mass on Earth weighs a little more than two pounds. **(220)**

L

leap year A year in which February has 29 days. This occurs every 4 years. **(191)**

2008   2012   2016
leap years

left **(300)**

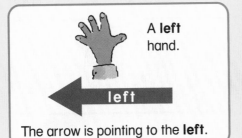

A **left** hand.

left

The arrow is pointing to the **left**.

length The distance from one point to another.
(209, 310, 311)

less than **(27)**

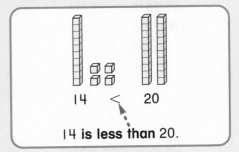

14 < 20

14 **is less than** 20.

likely Probably will happen. **(246)**

line A **line** is straight. It has no beginning and no end. **(194)**

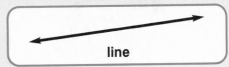

line

line of symmetry If a figure has a **line of symmetry**, you can fold the figure on the line and the two halves will match exactly. **(198)**

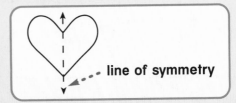

line of symmetry

line segment Part of a line. A **line segment** has a beginning and an end.

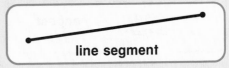

line segment

line symmetry If a figure can be folded so that the two halves match exactly, the figure has **line symmetry**. (198)

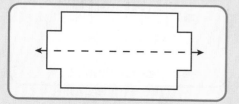

liter (L) A metric unit of capacity that is a little more than a quart. (224)

longer A word used when comparing the length of two objects. (208)

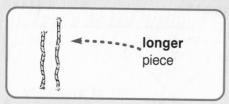

longer piece

longest A word used when ordering three or more objects by length. (208)

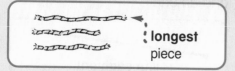

longest piece

M

mass A measure of the amount of matter in an object. **Mass** is different from *weight* because it does not change when the force of gravity changes. Kilograms and grams are units of **mass** in the metric system. Units of weight in the metric system are not used very often. (220, 311)

mental math Computing in your head. (114)

$$20 + 40 = 60$$

meter (m) A metric unit of length equal to 100 centimeters. A **meter** is a little longer than a yard. (212)

midnight 12 o'clock, the time between evening and morning. (182)

milliliter (mL) A metric unit of capacity. About 10 drops equal 1 **milliliter**. (224)

1000 **milliliters** = 1 liter

minute (min) A unit of time equal to 60 seconds. **(181)**

minute hand **(180)**

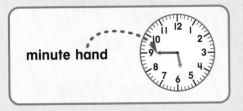

minute hand

missing addend **(84)**

$$7 + \blacksquare = 13$$

missing addend

money Coins and bills used to pay for things. **(164)**

multiplication The operation of combining equal groups to find a total. **(92)**

5 groups of 2
$5 \times 2 = 10$

multiply Combine equal groups. **(92)**

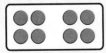

2 groups of 4
$2 \times 4 = 8$

Multiplying by 1 Property The product of any number and 1 is that number. **(263)**

$$1 \times 5 = 5$$

 N

nickel A coin worth 5 cents. **(164)**

a nickel

noon 12 o'clock, the time between morning and afternoon. **(182)**

not congruent Not having the same shape and the same size. **(200)**

same size, *not* same shape

not equal Not the same amount. **(259)**

$$4 \neq 5$$

4 is not equal to 5.

number sentence An equation or a comparison. **(258)**

$$3 + 4 = 7$$
$$8 - 2 = 6$$
$$7 > 6$$

number sentences

numeral A symbol used to stand for a number. **(298)**

1   53   468   **numerals**

numerator In a fraction, the number of equal parts talked about. The top number in a fraction. **(42)**

$$\frac{2}{3} \blacktriangleleft \text{-- -} \textbf{numerator}$$
(number of red parts)

octagon A figure with 8 straight sides. **(197)**

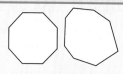

octagons

odd numbers Numbers that are *not* even. **(36)**

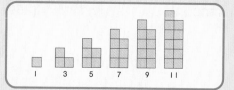

one fourth One of 4 equal parts. **(46)**

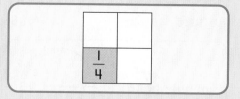

one half One of 2 equal parts. **(43)**

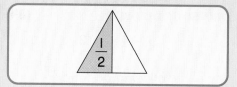

one hundred 10 tens. **(16)**

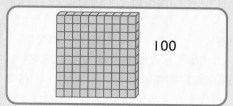
100

one third One of 3 equal parts. **(44)**

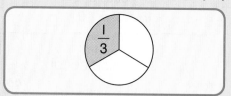

one thousand 10 hundreds. **(24)**

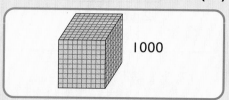
1000

open figure A figure that is not closed. **(197)**

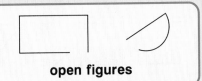

open figures

Order Property of Addition You can add two addends in any order and the sum will be the same. **(262)**

$$2 + 4 = 6$$
$$4 + 2 = 6$$

Order Property of Multiplication You can multiply two factors in any order and the product will be the same. **(263)**

$$3 \times 2 = 6$$
$$2 \times 3 = 6$$

ordered pair Two numbers that tell where a point is. **(264)**

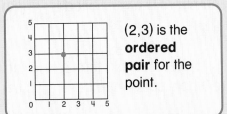

(2,3) is the **ordered pair** for the point.

ounce (oz) A customary unit of weight. **(218)**

16 **ounces** = 1 pound

pattern Something that changes in a regular way. **(254)**

Some **patterns** repeat.

Some **patterns** grow.

$$\begin{array}{ccccc} +1 & +2 & +3 & +4 & +5 \\ \hline 2 & 3 & 4 & 5 & 6 \end{array}$$

penny A coin worth 1 cent. **(164)**

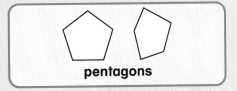

a **penny**

pentagon A figure with 5 straight sides. **(197)**

pentagons

perimeter The distance around a figure. **(214)**

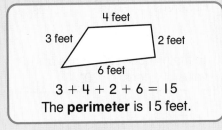

4 feet
3 feet
2 feet
6 feet

$$3 + 4 + 2 + 6 = 15$$
The **perimeter** is 15 feet.

pictograph (240)

Favorite Color of Shoes	
White	🧍🧍🧍🧍🧍🧍
Blue	🧍🧍🧍
Black	🧍🧍🧍🧍

Key 🧍 = 2 children

picture graph A graph that uses pictures to show data. Each picture stands for 1. (237)

Coins Saved					
Pennies	℗	℗	℗	℗	℗ ℗
Nickels	ⓝ	ⓝ	ⓝ	ⓝ	
Dimes	ⓓ	ⓓ			

pint (pt) A customary unit of capacity equal to 2 cups. (222)

place value The value a digit has because of its place in a number. (8)

Hundreds	Tens	Ones
▦	▊▊	🟥🟥🟥

Hundreds	Tens	Ones
1	2	3

In the number 123, 2 has a value of 2 tens, or 20.

plane figure A figure that lies completely on a flat surface. (196)

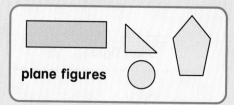

plane figures

pound (lb) A customary unit of weight equal to 16 ounces. (218)

prediction What you think will happen. (250)

I **predict** the next spin will be blue.

probability The chance of an event happening. (244)

product The result when numbers are multiplied. (93)

$4 \times 5 = 20$ product

property A rule that is true for a whole set of numbers. (262)

pyramid (204)

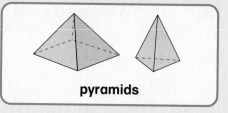

pyramids

Q

quadrilateral A figure with 4 straight sides. (197)

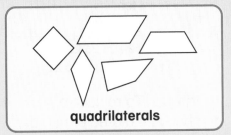

quadrilaterals

quart (qt) A customary unit of capacity equal to 4 cups. (222)

quarter A coin worth 25 cents. (164)

a quarter

R

real graph A graph made with real objects. (236)

rectangle A plane figure with 4 sides and 4 square corners. (196)

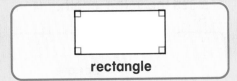

rectangle

rectangular prism (204)

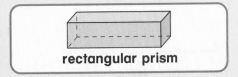

rectangular prism

regroup (126, 144)

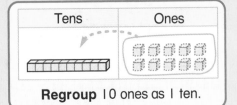

Regroup 10 ones as 1 ten.

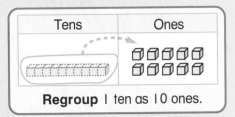

Regroup 1 ten as 10 ones.

remainder The amount left over after you divide into equal groups. (111)

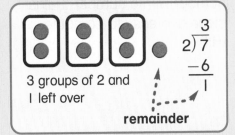

3 groups of 2 and 1 left over

$$2\overline{)7}$$
$$\begin{array}{r} 3 \\ 2\overline{)7} \\ -6 \\ \hline 1 \end{array}$$

remainder

rename *See regroup.*

rhombus A plane figure with all sides the same length. **(196)**

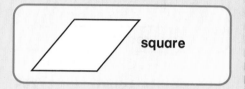

square

right **(300)**

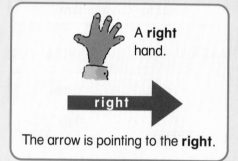

A **right** hand.

right

The arrow is pointing to the **right**.

right angle A square corner. **(195)**

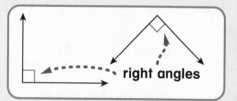

right angles

round a number To replace a number with a number that is close and easy to compute with. **(40)**

round down To replace a number with a lesser number that is easier to compute with. **(40)**

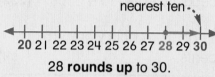

--- nearest ten

20 21 22 23 24 25 26 27 28 29 30

23 **rounds down** to 20.

round up To replace a number with a greater number that is easier to compute with. **(40)**

nearest ten -

20 21 22 23 24 25 26 27 28 29 30

28 **rounds up** to 30.

S

scale **(219)**

shorter A word used when comparing the length of two objects. **(208)**

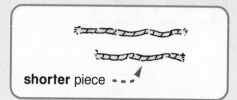

shorter piece - - -

shortest A word used when ordering three or more objects by length. **(208)**

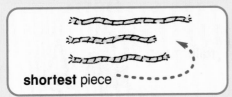

shortest piece

side **(197)**

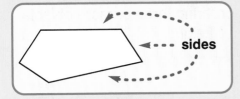

sides

skip count **(166)**

Skip counting by fives:

5   10   15   20   25

Skip counting by twos:

2   4   6   8   10

slide One way to move a figure. **(202)**

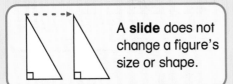

A **slide** does not change a figure's size or shape.

solid figure A figure that is not flat. **(204)**

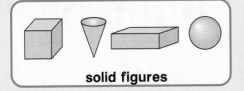

solid figures

sphere **(204)**

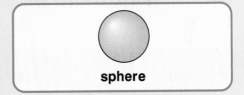

sphere

square A figure with 4 sides that are the same length and 4 right angles. **(196)**

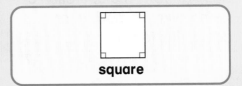

square

square centimeter A square with sides 1 centimeter long. **(216)**

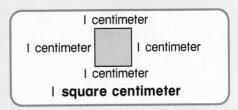

1 centimeter

1 centimeter   1 centimeter

1 centimeter

1 **square centimeter**

square unit A square with sides 1 unit long. **(216)**

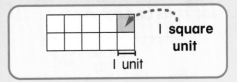

1 **square unit**

1 unit

statistics The mathematics of data. The numbers used to describe data. **(234)**

strategy A plan or method used to solve a problem or win a game. **(58)**

subtract Take away, remove, or compare. **(73, 76, 102)**

subtraction The operation of taking away, removing, or comparing to find a difference. **(72)**

sum The result when numbers are added. **(55)**

$5 + 7 = 12$ **sum**

survey A way to gather data by asking questions. **(234)**

symmetry *See line symmetry.*

table A way to organize data using rows and columns. **(235)**

Price List	
Size	Cost
small	10¢
medium	25¢
large	50¢

taller A word used when comparing the height of two objects. **(208)**

taller tree

tallest A word used when ordering three or more objects by height. **(208)**

tallest tree

tally chart **(234)**

Plants Sold		
Zinnia	̶H̶H̶T̶ ///	8
Mum	̶H̶H̶T̶ ̶H̶H̶T̶ //	12
Ivy	̶H̶H̶T̶ /	6
Violet	̶H̶H̶T̶ ̶H̶H̶T̶ ̶H̶H̶T̶ ////	19

tally marks **(234)**

/   //   ///   ////   ̶H̶H̶T̶   ̶H̶H̶T̶ /
1   2   3   4   5   6

Tally marks are recorded in groups of 5.

temperature A measure of how hot or cold something is. **(226)**

thermometer A tool used to measure temperature. **(226)**

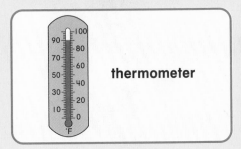

thermometer

thirds The parts you get when you divide something into 3 equal parts. **(44)**

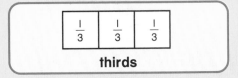

thirds

time Seconds, minutes, hours, days, months, years, and so on. **Time** is shown on a clock or calendar. **(180)**

trade *See regroup.*

triangle A figure with 3 straight sides. **(196)**

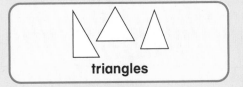

triangles

turn One way to move a figure. **(203)**

A **turn** does not change a figure's size or shape.

 U

unit A quantity used for measuring. **(209)**

unlikely Probably will *not* happen. **(246)**

 V

Venn diagram **(242)**

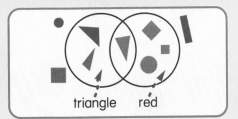

triangle   red

vertical line A line going straight up and down. **(194)**

vertical line

weight A measure of how heavy something is. **(218, 310)**

yard (yd) A customary unit of length equal to 36 inches, or 3 feet. **(210)**

Z

Zero Property of Multiplication The product of any number and zero is zero. **(263)**

$$5 \times 0 = 0$$

Index

Illustration Credits:
Creative art: Mike Gordon
Technical art: Bill SMITH STUDIO
Cover: Bill SMITH STUDIO

Symbol	Meaning	Example
+	plus	$5 + 2 = 7$
−	minus	$7 - 2 = 5$
×	multiplied by	$3 \times 4 = 12$
÷	divided by	$8 \div 2 = 4$
$\overline{)}$	divided by	$2\overline{)8}$ with 4 above
=	is equal to	$3 + 7 = 10$
≠	is not equal to	$3 + 7 \neq 9$
<	is less than	$5 < 8$
>	is greater than	$5 > 2$
¢	cents	15¢
$	dollars	\$1.50
°F	degrees Fahrenheit	32°F
°C	degrees Celsius	50°C
right angle	right angle	